Lyme disease

Alain L. Fymat

Lyme disease

ALSO FROM THE SAME AUTHOR:

Lectures on the Physics of Radiology

A Guide for the Teaching of Clinical Radiological Physics
(with A.G.H. McCollough *et al.*)

Management and Treatment of Glioblastoma

From the Heart to the Brain

The Odyssey of Humanity's Diseases – Volumes 1, 2, 3

Alzhei...Who?

Parkin ...ss oo..nn

… and other books

Lyme disease

**The dreadful invader, evader, and imitator
... and what you can do about it**

Alain L. Fymat

International Institute of Medicine & Science Press

Lyme Disease
Copyright © 2020 by Alain L Fymat

All rights reserved. No part of this publication may be reproduced, distributed, or transmitted in any form or by any means, including photocopying, recording, or other electronic or mechanical methods, without the prior written permission of the author, except in the case of brief quotations embodied in critical reviews and certain other non-commercial uses permitted by copyright law.

Tellwell Talent
www.tellwell.ca

ISBN
978-0-2288-3198-3 (Hardcover)
978-0-2288-3199-0 (Paperback)

DISCLAIMER: The information in this book has been compiled by way of general guidance concerning the specific subjects addressed. It is not a substitute, and should not be relied on, for medical, healthcare, pharmaceutical or other professional advice on specific circumstances, and in specific locations. Please consult your General Practitioner or Physician Specialist before starting, changing, or stopping any medical treatment. So far as the author is aware, the information given is correct and current as of December 2019. Practice, laws, and regulations all change and the reader should obtain up-to-date professional advice on any such issue. As far as the laws allow, the author disclaims any liability arising directly or indirectly from the use or misuse of the information contained in this book.

*To Robert who may have been undiagnosed, misdiagnosed or over-treated
and
to all those who, unknowingly having lived in or visited a harmful environment, have the misfortune of being afflicted by this dreadful disease.*

Contents

Chapter		Page
	Preface	7
	Introduction	9
1	The town and its namesake disease.	14
2	But,... what is LD?.	18
3	What are CLD... and PLDS?.	36
4	Signs and symptoms of LD.	48
5	Getting the correct LD diagnosis	59
6	Clinical practice guidelines	74
7	Can CLD be treated... and how?.	85
8	Is there a homeopathic treatment and its co-infections?.	93
9	Is there a vaccine against LD?.	99
10	Prevention and prognosis	110
11	Is STARI the same as Lyme?.	121
12	What you need to know about Lyme neuroborreliosis	125
13	What you need to know about Lyme carditis.	138
14	Pathogens... in the brain?.	143
15	Epidemiology of Lyme.	152
16	Frequently asked questions.	160
17	What about the other diseases transmitted by the Lyme vector?.	170
18	What about other tick-borne diseases in the U.S.?.	182
19	What can you do beyond seeking treatment?.	192
20	The voice of the patient - Living with Lyme.	206
21	Any recent out-of-the-box research developments?.	215
22	Latest news in the world of Lyme.	223
23	Controversies and challenges in treating Lyme and other tick-borne diseases.	229
References		241
Illustrations.		252
Tables		253
Sidebars.		254
Glossary.		255
Abbreviations		263
Subject index.		266
Author index.		274
About the Author.		277

Preface

Lyme disease, also known as Lyme *borreliosis*, is a tick-borne zoonotic disease that affects humans and animals. It is a bacterial, vector-borne infection due to *borrelia* (a helicoid spirochete) that is transmitted by bites from the *Ixodes* tick. *Borrelia* includes a dozen of species (36 as of the end of 2018, but others may still be discovered) that were named in their entirety after the French bacteriologist Amedee Borrel (1867-1936). Whereas *borrelia burgdorferi* is the best known of them for being primarily responsible for Lyme disease in the United States, dozens of other *borrelias* had been identified in Europe and registered (under different names) since the beginning of the 20th century or long before the discovery of *borrelia burgdorferi* in the mid-1970s. The ticks are found in temperate forested regions of North America, Europe, and Asia, generally at elevations less than 1300 meters.

Lyme disease was identified in the U.S. in 1975 after a mysterious outbreak of what appeared to be juvenile rheumatoid arthritis (also called juvenile idiopathic arthritis) in children who lived in Lyme and Old Lyme, Connecticut. However, there is no relationship between juvenile rheumatoid arthritis and Lyme disease for which it was originally (mis)named. Until then, Rocky Mountain spotted fever had been the main concern.

Several ancient lines, largely distributed across the globe, including in North America, have recently been interpreted as an epidemic development. However, their diversity and wide distribution is incompatible with the emerging hypothesis of a recent genetic modification, let alone a genetic manipulation. According to this hypothesis, Lyme had been utilized as a biological weapon, as current discussions taking place in the U.S. since 2019 would suggest. Rather, and more simply, the emergence of Lyme as a concerning disease may result from anthropometric perturbations of the ecosystem and climate changes that have favored the widespread multiplication of ticks or/and bacteria and the corresponding increasing number of human infections.

Lyme disease is a complex disease, which is often difficult to diagnose and treat effectively. Like other zoonoses, it is transmitted by a vector that picks up the pathogen during a blood meal from a vertebrate host. In the eastern and central United States, the spirochete bacterium *borrelia burgdorferi* infects black-legged ticks, *Ixodes scapularis*, which feed on a wide variety of birds, lizards, and mammals including mice, deers, and humans. Since human risk is a function of the prevalence of infection among vectors, outbreak prevention depends in part on understanding what controls infection rates among the agents of transmission. Typical symptoms, which may appear in whole or in part, include fever, headache, fatigue, and a characteristic skin rash called *erythema migrans*. Symptoms may vary depending on the specific type of *borrelia*. In North America, the principal species is *borrelia burgdorferi sensu stricto*, which is particularly likely to also cause arthritis. In contrast, the European species *borrelia garinii* and *borrelia afzelii* are more often associated with neurological and chronic dermatologic manifestations, respectively. Approximately 85,000 cases of Lyme disease are reported annually in Europe whereas, in the U.S., recent studies suggest that approximately 300,000 people are diagnosed each year. If left untreated, infection can spread to the joints, the heart, and the nervous system. Fortunately, most cases of Lyme disease can be treated successfully with a few weeks of antibiotics.

Early in the history of Lyme in the U.S., many patients reported that commonly prescribed antibiotic regimens failed to restore their health, leaving them with significant long-term quality of life impairments. Still, despite improvements in diagnostic testing and public awareness of Lyme disease, the reported cases have increased over the past decade to approximate 300,000 per year. These limitations and the failed public acceptance have led to the demise of a human vaccine. In response to a growing list of patient concerns, a group of insightful physicians and scientists decided to take action to better alleviate patient suffering. They formed a professional society (the International Lyme and Associated Diseases Society, ILADS) to provide support for researchers, providers, and patients in the pursuit of more effective treatment of tick-borne illnesses. ILADS disagrees with mainstream consensus medical views on Lyme disease, leading it to adopt and publish alternative diagnostic criteria and treatment guidelines. Further, it sustains the controversy as to the existence of chronic Lyme disease by advocating for long-term antibiotic treatment. However, the existence of persistent *borrelia* infection is presently not supported by high-quality clinical evidence. In addition, the mainstream medical profession has deemed the use of long-term antibiotics as dangerous and contraindicated.

Lyme and other tick-borne illnesses remain poorly understood so that current guidelines addressing the illness in surveillance case terms do not serve patients well. Further, challenges and controversies about the disease devolve from its existence in a setting of incomplete scientific evidence, including a lack of validated direct testing methods which can be applied across all stages of the disease to accurately distinguish infected from uninfected patients. Further, serologic testing, often held as a gold standard, has significant performance limitations.

In this book, I attempt to explain the many challenges and controversies faced by patients with tick-borne illnesses, including those over terminology, diagnosis, and treatment. I also try to describe ways patients and, in some cases also their caregivers, can do something about it in addition to seeking medical treatment.

Introduction

Lyme disease, also known as Lyme *borreliosis*, is a tick-borne zoonosis caused by infection with the spirochete *borrelia burgdorferi*. There are three varieties depending on the pathogen transmitted and the carrier tick. The number of annually reported cases in the United States has increased about 25-fold since national surveillance began in 1982. The disease is primarily localized to States in the northeastern, mid-Atlantic, and upper north-central regions, and to several areas in northwestern California.

Lyme disease is a multi-stage, multi-system inflammatory illness. In its early stages, it can be treated successfully with oral antibiotics; however, untreated or inadequately treated, it can progress to late-stage complications requiring more intensive therapy. The first line of defense against the disease and other tick-borne illnesses is avoidance of tick-infested habitats, use of personal protective measures (e.g., repellents and protective clothing), and checking for and removing attached ticks. Early diagnosis and treatment are effective in preventing late-stage complications. However, even after successful treatment with antibiotics and resolution of their clinical symptoms, some patients can still have the bacteria in their blood for decades. When the key symptoms are kept at bay, many believe that this is a normal humoral immune response and not a continuation of active chronic infection.

Most often, Lyme disease is evidenced by a characteristic rash (*erythema migrans*) accompanied by nonspecific symptoms (e.g., arthralgia, fatigue, fever, headache, malaise, and myalgia). The incubation period from infection to onset of *erythema migrans* is typically 7-14 days but can be as short as 3 days or as long as 30 days. Some infected persons have no recognized illness (i.e., their infection is asymptomatic as determined by serologic testing), or they manifest only nonspecific symptoms (e.g., fatigue, fever, headache, and myalgia).

Lyme disease spirochetes disseminate from the site of inoculation by cutaneous, lymphatic, and blood-borne routes. Through a number of mechanisms, the microbe is adept at countering, distracting, and evading the host's adaptive immune system. When *borrelia* finally settles down and multiplies, it secretes a protein mesh (called a "biofilm") which shields it and its progeny from direct exposure to many antibiotics and other compounds. While antibiotics cannot easily pass the blood-brain barrier (that physical and physiological gate that separates the circulating blood from the brain), *borrelia* and other infections can cross it, attacking and infecting the brain and the central nervous system. The risk of certain complications of the condition may be influenced by inherited genetic factors but the disease itself cannot be inherited. The signs of "early-disseminated" infection usually occur from days to weeks after the appearance of a solitary *erythema migrans* lesion. In addition to multiple or secondary *erythema migrans* lesions, early-disseminated infection can be manifested as a disease of the nervous system, the musculoskeletal system, or the heart. Early neurologic manifestations include lymphocytic meningitis; cranial neuropathy, especially facial nerve palsy; and radiculoneuritis. Musculoskeletal manifestations can include migratory joint and muscle pains with or without objective signs of joint swelling. Cardiac manifestations are rare but can include myocarditis and transient atrioventricular block of varying degree.

Borrelia burgdorferi infection in the untreated or inadequately treated patient can progress to "late-disseminated" disease from weeks to months after infection. The most common objective manifestation in this case is intermittent swelling and pain of one or some joints, usually the large, weight-bearing joints (e.g., the knee). Some patients experience chronic axonal polyneuropathy, or encephalopathy, the latter usually manifested by cognitive disorders, sleep disturbance, fatigue, and personality changes. Infrequently, Lyme disease morbidity can be severe, chronic, and disabling. An ill-defined *post-Lyme disease syndrome* occurs in some persons after treatment. Fortunately, Lyme disease is rarely, if ever, fatal.

In this volume, I will provide a detailed review of Lyme and associated tick-borne diseases, mostly those encountered in the U.S. Chapters 2 to 4 will define Lyme disease, "chronic Lyme disease", and "post-treatment Lyme disease syndrome". While *borrelia* is responsible for causing primary Lyme disease, it is usually not the only factor. When one or several ticks perhaps from different species do bite, various bacteria and parasites also enter the wound causing other opportunistic infections. The overall disease caused by both the primary and secondary infections is referred to as "chronic Lyme disease" or "chronic Lyme syndrome" or "late Lyme". Many of the associated infections complicate treatment and increase the risk for autoimmune diseases depending on the patient's genetic background. They can also be a precursor to various cancers and even neurodegenerative diseases. Treating these infections in their totality as quickly as possible is essential to full recovery and to achieve health. Blood tests often miss this type of Lyme since the acute phase has passed. Chronic Lyme can cause a significant burden on patients, linger for months or even years, typically worsen over time in the absence of proper treatment, and ultimately lead to debilitating symptoms.

Because of the confusion in how the term chronic Lyme disease is employed, and the lack of a clearly defined clinical definition, many experts in the field do not support its use so much so that it is not an officially recognized diagnosis. By contrast, the International Lyme and Associated Diseases Association advocates for it, providing corresponding diagnosis and therapeutics guidelines.

Post-treatment Lyme disease syndrome is purportedly caused by persistent infection, but this is not believed to be true because we are unable to detect infectious organisms after standard treatment. Rather, it describes the continued ongoing resistance to antibiotics and continued symptoms.

The several primary and secondary infections release a multitude of endotoxins, neurotoxins, and biotoxins that confuse the immune system and cause it to potentially attack its own cells, resulting in autoimmune-like symptoms and chronic inflammation throughout the brain and the body. It is the combination of these multiple infections with other complications that contribute to the patients' debilitating set of physical and neurological symptoms. Correctly identifying and vigorously treating them is the key to getting the correct diagnosis and producing lasting results against chronic Lyme disease, as will be discussed in Chapter 5. The several diagnostic tests employed will be described and analyzed including ELISA, Western blot, dark-field microscopy, Ispot, polymerase chain reaction, genomic testing, and next generation sequencing. This latter test, in particular, is the most accurate of all and may allow greater sensitivity than the other tests. Other neurologic and cardiologic tests will also be discussed. Lyme patients with neurological symptoms are often misdiagnosed with one or more neurological diseases. Late-stage Lyme disease may also be misdiagnosed as multiple sclerosis, rheumatoid arthritis, fibromyalgia, chronic fatigue syndrome, lupus, Crohn's disease, HIV or other autoimmune and neurodegenerative diseases.

Clinical practice guidelines will be set forth in Chapter 6. They consist of a two-test methodology using a sensitive enzyme immunoassay or immunofluorescence assay as a first test, followed by a western immunoblot assay for specimens yielding positive or equivocal results. Unfortunately, this approach relies too much on old technology, does not account for newer technological developments (such as genomic testing), and does not espouse principles of modern integrative and personalized medicine. It also often provides false negatives, resulting in a monumental failure to provide the needed critical early detection and treatment. Thus, most Lyme disease tests are designed to detect antibodies made by the body in response to infection. However, antibodies can take several weeks to develop so patients may test negative if infected only recently. Further, antibodies normally persist in the blood for months or even years after the infection is gone, therefore, the test cannot be used to determine cure. In addition, other tick-borne associated diseases and some viral, bacterial, or autoimmune diseases can result in false positive test results.

Chapter 7 will be particularly concerned with the treatment of chronic Lyme disease and post-treatment Lyme disease syndrome when patients often relapse. The conventional wisdom is that long-term antibiotic treatment can be associated with serious, sometimes deadly complications and can be dangerous. However, the International Lyme and Associated Diseases Society rather advocates for such a long term use. Why some patients experience the syndrome is not known and there is no known treatment for it. Fortunately, in the latter instance, patients usually get better over time, but it can take many months to feel completely well.

A homeopathic alternative to chronic Lyme treatment will be described in Chapter 8. It has been advocated by some health care providers. I do not necessarily condone it but present it here for completeness of the information I provide. No less than 21 different botanicals have been identified (there may be more), each one offering unique properties and several of them acting synergistically to allegedly help lower inflammation, reduce joint pain, fight free radical damage, break up biofilm, decrease viral load, regulate the immune system, purge parasites, support detoxification, and combat Lyme bacteria to beat chronic Lyme. Nonetheless, before resorting to any homeopathic treatment, I strongly recommend that you first consult with your physician and the FDA to ascertain the treatment's safety and efficacy, alleged benefits, and legal distribution.

Chapter 9 will relate the failed history of a vaccine against Lyme that was marketed in the U.S. between 1998 and 2002. Currently, there is no vaccine against LD and those people who were vaccinated are probably no longer protected as the protection diminishes over time. There are, however, several possible avenues regarding new vaccine developments that will be discussed. A vigorous initiative is here needed for the targeted prevention of tick-borne diseases.

As will be discussed in Chapter 10, tick bites may be prevented by avoiding or reducing time spent in likely tick habitats and taking precautions while in and when getting out of such habitats. Several precautionary measures will be described, which can be taken to repel or/and kill ticks.

There is more to say about rashes without them being Lyme disease as will be emphasized in Chapter 11. This will be exemplified by the southern tick-associated rash illness (STARI) following bites by the lone star tick *Amblyomma americanum*. It manifests itself by a rash similar to Lyme's rash. Symptoms resolve following treatment with an oral antibiotic (doxycycline), but it is not known whether antibiotic treatment is necessary or beneficial or merely speeds recovery.

Chapter 12 will succinctly discuss Lyme *neuroborrelliosis*, a manifestation of Lyme as a disorder of the central nervous system. Common symptoms include headache, sleep disturbance, and intracranial pressure such as papilledema. Less common symptoms include meningitis, myelitis ataxia, and chorea. Confounding other diseases include Alzheimer's disease; acute disseminated encephalomyelitis; viral meningitis; multiple sclerosis; Bell's palsy; and amyotrophic lateral sclerosis. Chapter 13 will likewise discuss Lyme carditis, an uncommon manifestation of "early-disseminated" Lyme disease, which occurs when Lyme disease bacteria enter the tissues of the heart, interfere with the normal movement of electrical signals from the heart's upper to lower chambers, and result in an "atrioventricular heart block". While usually treated by a 2-4 week course of an antibiotic (doxycycline or ceftriaxone), serious cases of heart block necessitate hospitalization, IV antibiotics, and in some cases even the placement of a temporary pacemaker.

Pathogens in the brain (cytomegalovirus, bacteria, viruses, fungi, and other microbes) will be discussed in Chapter 14 as they are found in the brains of patients with neurodegenerative diseases like Parkinson's, Alzheimer's, and other diseases that were not thought to be infectious. Brain pathogens of interest will be those for Lyme, *Ehrlichia*, *Babesia*, and *Bartonella*. However, while viral (and other pathogen) infections can damage the brain, there is currently no definitive study demonstrating that a virus (or other pathogen) can cause Parkinson's disease or Alzheimer's disease or a number of other neurodegenerative diseases.

The epidemiology of Lyme will be briefly summarized and illustrated in Chapter 15 by geographical distribution, gender, and age group. Chapter 16 will answer a list of frequently asked questions. Other important co-infective and confounding diseases transmitted by the Lyme tick vector will be surveyed in Chapters 17 and 18 including *anaplasmosis, ehrlichiosis, babesiosis,* and *Powassan virus disease*. It behooves everyone to be aware of the degree of risk of getting any of the above infectious diseases when living in, or visiting, any of the States of the Union.

I am suggesting in Chapter 19 a number of actions you can take in addition to seeking treatment. I will also list the several private, professional, national, and international organizations (13 or more) that offer invaluable advocating and supporting services. With the advice and consent of your physician, you may also participate in clinical trials whether in the U.S. or abroad (as of March 2020, 26 of them are recruiting volunteers).

By giving voices to patients, I will also outline in Chapter 20 how to live with Lyme, benefiting from the lessons learned in these and other cases. I will also review there the case of peripheral neuropathy and discuss its treatment options.

Lyme disease

Recent out-of-the-box research developments have taken place. These will be summarized in Chapter 21 including, among other findings, the following: repeated Lyme infection has an immune priming effect; a new cognitive fingerprint has been identified for post-treatment Lyme disease; Lyme can persist in the human body not only in the spirochetal but also in the antibiotic-resistant biofilm form, even after long-term antibiotic treatment; more effective antimicrobial drugs than Vancomycin, used early in the infection, may prevent or reduce the occurrence of persisting infection; drug-tolerant persister cells in Lyme can be eradicated by an appropriate combination of antibiotics; the combination (daptomycin + mitomycin + ceftriaxone) can kill all persister cells and eradicate all live bacteria; medical marijuana has merit as a second- or third-line of treatment for peripheral neuropathy; viral gene therapy blocks peripheral nerve damage in mice by preventing axon destruction; stem cell treatment is still a hope in neuropathic pain treatment when conducted in a reputable medical research institution; scrambler therapy can send a "non-pain" signal along the same pain fibers that are sending the "pain" stimulus; there is no evidence that the flu shot will make neuropathy symptoms worse unless the person had a vaccine-induced neuropathy associated with Guillain-Barre Syndrome; *Borrelia burgdorferi* persists in the gastrointestinal tract of children and adolescents with Lyme disease even after antibiotic treatment. And children with chronic Lyme disease have cognitive and psychiatric disturbances that may result in psychosocial and academic impairments.

The latest news in the world of Lyme will be summarized in Chapter 22. Lastly, Chapter 23 will review the consensus of major U.S. medical authorities relating to chronic Lyme disease and post-Lyme syndrome, and analyze the controversies and challenges in the diagnosis, treatment, and other challenges of Lyme and other tick-borne diseases as advocated by the International Lyme and Associated Diseases Society.

The evidence regarding Lyme and other tick-borne associated diseases is evolving. Unfortunately, many in the greater medical community may not be sufficiently versed in the underlying scientific evidence, giving rise to misconceptions about the disease.

For the reader's convenience, each Chapter ends with a take-home points section to summarize what we have learned from it. Sidebars are also provided at the ends of Chapters 1, 3, 6, 9, 10, 12, 14, and 19 for those readers interested in the more specialized aspects they describe.

1
Lyme - The town and its namesake disease

Lyme is a historic, quaint, little coastal town in New London County, Connecticut, United States. It is located on the east bank of the Connecticut River at its confluence with the Long Island Sound, across the river from Old Saybrook on the west bank. It has a venerable history dating back to at least 1665 when the territory of the Saybrook Colony east of the Connecticut River was set off from Saybrook (now known as Deep River). In 1667, the Connecticut General Court officially recognized the Saybrook plantation as the town of Lyme, including present-day Lyme, Old Lyme, and the western part of East Lyme. South Lyme was later incorporated from Lyme in 1855, then renamed Old Lyme in 1857 because it was not only one of the early settlement areas of the Puritan Saybrook Colony in the 17th century but also the early town center of Lyme. It contains the oldest settled portion of the "Lymes". Its neighbor to the north is the town of Lyme. The eastern part of Lyme (bordering the town of Waterford) separated from Lyme and became East Lyme in 1823, and the southern portion of Lyme along Long Island Sound separated as South Lyme in 1855 and renamed Old Lyme in 1857. Numerous examples of Colonial and Federal architecture can be found throughout the town. The picturesque Old Lyme historic district is pictured in Figure 1.1 with the cemetery containing the graves of the original settlers. Figure 1.2 depicts one of the several bucolic pictures of Lyme while Figure 1.3 shows the railroad bridge over the Four-Mile River's mouth, which connects East Lyme to Old Lyme's eastern shore, facing Old Lyme's Seaview Road shore.

Old Lyme has long been a popular summer resort and an artists' colony. It contains several villages, including Black Hall, Laysville, Lyme, Soundview, and South Lyme. It houses the Florence Griswold Museum (including the Florence Griswold House), the Lyme Academy College of Fine Arts, and the Lyme Arts Association.

According to the 2010 United States Census Bureau, its population was 2,406 inhabitants having grown from 2016 people at the previous 2000 census (an increase of less than 1.5% in 10 years). Figure 1.4 is the official seal of Old Lyme, Connecticut.

The place name "Lyme" derives from Lyme Regis, a small port on the coast of West Dorset in Southern England from which it is believed the early settlers migrated in the 17th century. Styled "The Pearl of Dorset", it lies at Lyme Bay on the English Channel coast at the Dorset-Devon border.

Figure 1.1 – The Old Lyme historic district

Source: www.townlyme.org

Figure 1.2 – View of Lyme's countryside

Figure 1.3 - Railroad Bridge over the Four-Mile River's mouth

Nothing predicted that quiet Lyme and its neighboring town Old Lyme would become the namesake for a devastating disease. In the U.S., Lyme disease was discovered in 1975 after a mysterious outbreak of what appeared to be juvenile rheumatoid arthritis in children who lived there. Also called juvenile idiopathic arthritis, the disease is the most common form of arthritis in children and adolescents. (Juvenile, in this context, refers to an onset before age 16 while idiopathic refers to a condition with no defined cause, and arthritis is the inflammation of the synovium of a joint.) It is an autoimmune, non-infective, inflammatory joint disease of more than 6 weeks duration in children less than 16 years of age. The disease commonly occurs in children from the ages of 1 to 6, but it may develop as late as 15 years of age. It is a subset of arthritis seen in childhood, which may be transient and self-limited or chronic. It differs significantly from arthritis commonly seen in adults (osteoarthritis, rheumatoid arthritis), and other types of arthritis present in childhood that are chronic conditions such as, for example, psoriatic arthritis and ankylosing spondylitis. It affects about one in 1,000 children in any given year, with about one in 10,000 having a more severe form.

However, there is no relationship between that outbreak of juvenile rheumatoid arthritis and Lyme disease for which it was originally mistaken. But what is Lyme disease? This will the subject of our next Chapter.

Figure 1.4 – Official seal of Old Lyme, Connecticut

Take-home points

- In the U.S., Lyme disease was discovered in 1975 after a mysterious outbreak of what appeared to be juvenile rheumatoid arthritis (also called juvenile idiopathic arthritis) in children who lived in Lyme and Old Lyme, Connecticut. It was discovered much earlier in Europe.

- There is no relationship between juvenile rheumatoid arthritis and Lyme disease for which it was originally named.

2

But... what is LD?

In the U.S., Lyme disease (LD) was diagnosed as a separate condition for the first time in 1975 in Old Lyme, Connecticut. It is an infectious disease that, as I indicated earlier, was originally mistaken for juvenile rheumatoid arthritis (also called juvenile idiopathic arthritis). It is the most commonly occurring vector-borne infection spread by ticks in the Northern Hemisphere and the sixth most commonly reported notifiable infectious disease. Annually, it is estimated to affect about 300,000 people in the United States and 85,000 people in Europe.

A brief history of LD

Borrelioses have long existed! The existence and evolution of *Borrelia burgdorferi* in the U.S. have been reconstructed, suggesting it dates back more than 60,000 years, long before the arrival of humans in the Americas. The first known man (named Otzi) having the disease was identified about 5,300 years ago from the residual bacterial DNA recovered from his cadaver. However, the manifestations of the disease had been described, *albeit* sparsely, in Europe since the end of the 19th century. Thus, in 1883, a German physician in Breslau, Alfred Buchwald, described it as a skin anomaly which resembled what we today would call *atrophic chronic acrodermatitis* (ACA) but did not link it to a tick bite. That *borreliosis* was different from Lyme disease proper as the dominant *borrelia* was different from those found in Eurasia or North America. In 1909, a Swedish dermatologist, Arvid Afzelius had noted the apparition of a ring-like lesion following a bite by the *Ixodes* tick and called it *erythema migrans*. A few years thereafter, an Austrian dermatologist, called it a "migrant chronic erythema". In 1922, French physicians Garin and Bujadoux associated this dermatologic lesion with an incidence of paralysis "...*more or less serious, at times deadly, consecutive to tick bites (Ixodes hexagonus)*". They attributed this lymphocytic meningo-radiculopathy (LMR, a combination of meningitis and polyneuritis) to a "virus" to be found not in the blood but in nervous tissues. Subsequently, in 1934, in Germany, the *erythema migrans* was described in association with arthritis. In 1944, again in Germany, Alfred Bannwarth also associated the arthritis with LMR. The corresponding disease was therefore named the Garin-Boujadoux-Bannwarth syndrome (GBBS). Later, in 1951, the beneficial effects of penicillin pointed to the bacterial origin of these other expressions of the disease. As related in Chapter 1 and above, it is only in 1975, in Lyme. Connecticut, U.S.A. that the mysterious outbreak of juvenile rheumatoid arthritis (or juvenile idiopathic arthritis) in children was recognized and misnamed Lyme disease.

The initial culprit

There are two different LD pathogens that will later be discussed within the context of Table 2.1 below, namely (1) *Borrelia burgdorferi*, which can be transmitted by either the tick *Ixodes scapularis* or/and the tick *Ixodes pacificus* and (2) *Borrelia mayonii* transmitted by the tick *Ixodes scapularis* only.

Figure 2.1 – Dr. W. Burgdorfer in 1978
(Born in Basel, Switzerland 27 June 1925, died November 1, 2014 in Hamilton, Montana, U.S.A.)

Borrelia burgdorferi was first described in 1981 by Wilhelm "Billy" Burgdorfer (1925-2014), an American scientist born and educated in Basel, Switzerland, who was considered an international leader in the field of medical entomology (Figure 2.1). He gained worldwide recognition for his 1982 discovery of a tick-borne spirochete (a cousin of the syphilis family of diseases) as the long sought-after cause of LD and related disorders in the U.S. and Europe. The pathogen was named *Borrelia burgdorferi* in his honor. Dr. Burgdorfer was interested in the interactions between animal and human disease agents and their transmitting arthropod vectors, particularly ticks, fleas, and mosquitoes. His research contributions are published in more than 225 papers and books. They cover a wide field of investigations including those on relapsing fever, plague, tularemia, Colorado tick fever, Rocky Mountain spotted fever, and other bacterial and viral diseases.

Figure 2.2 - *Borrelia* bacteria, the principal causative agents of Lyme disease

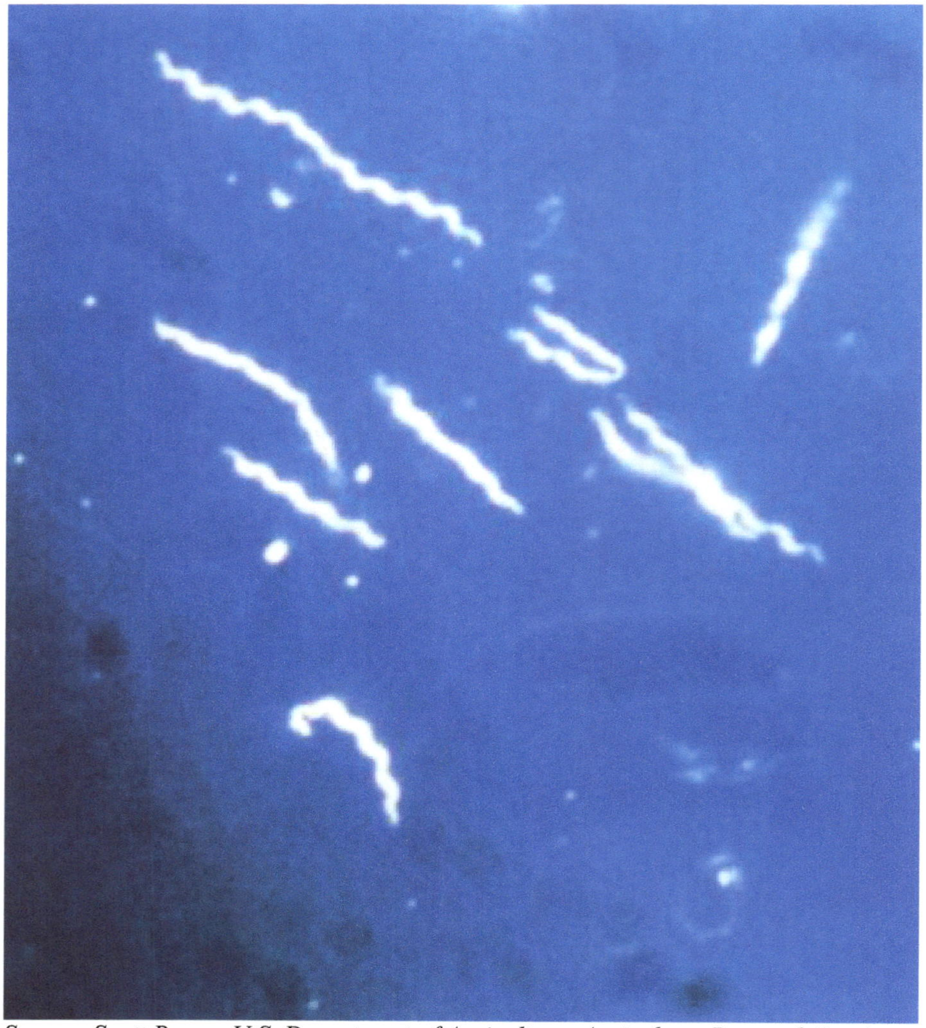

Source: Scott Bauer, U.S. Department of Agriculture, Agriculture Research Service

Figure 2.2 shows a population of the cork-screwed *Borrelia* spirochetes of different lengths and Figure 2.3 shows it stained with wheat germ agglutinin.

Figure 2.3 - Live *Borrelia burgdorferi sensu stricto* stained with wheat germ agglutinin (Alexa Fluor®)

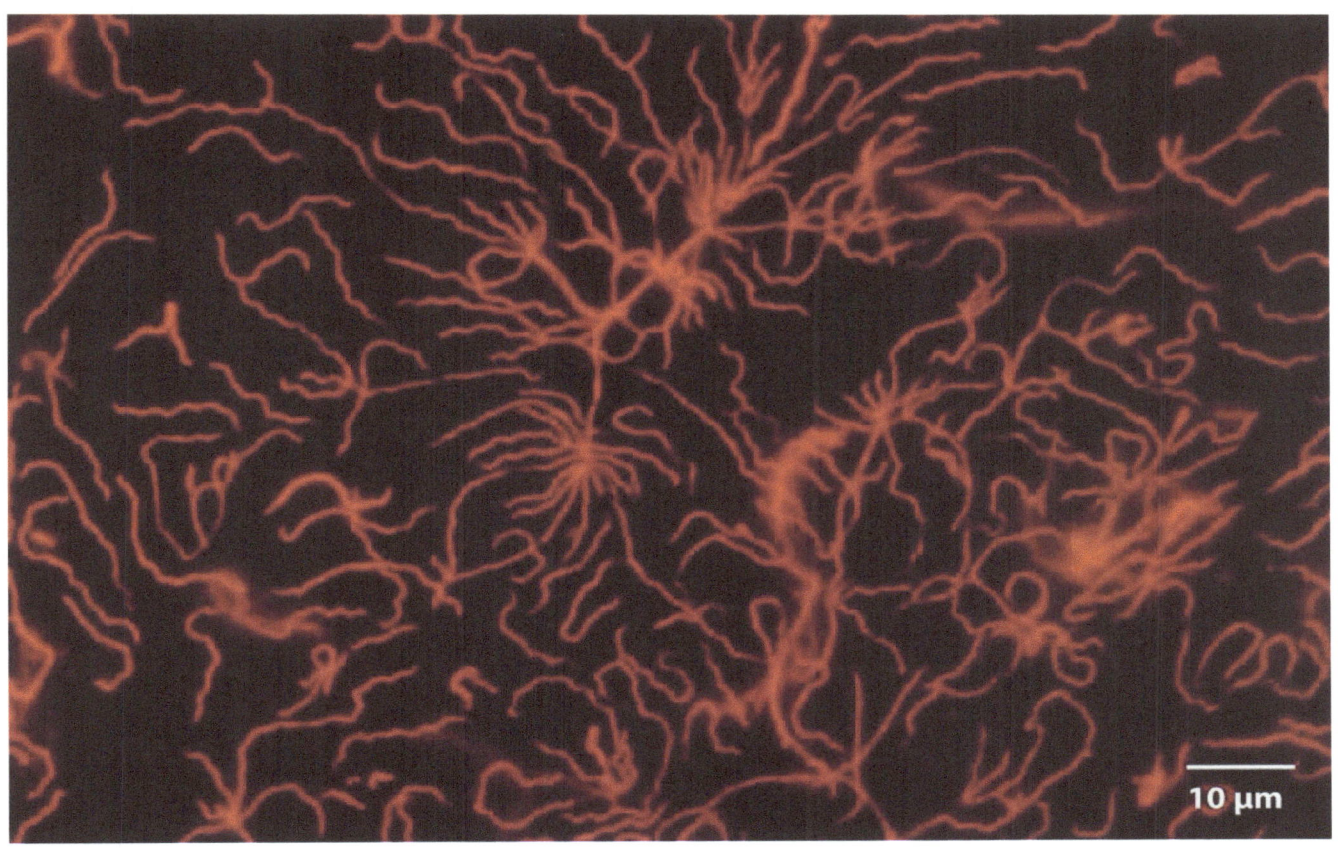

Clinical manifestations of the disease

Most often, LD is evidenced by a characteristic rash (called *erythema migrans*, EM) that is accompanied by nonspecific symptoms (e.g., fever, malaise, fatigue, headache, myalgia, and arthralgia). The incubation period from infection to onset of EM is typically 7-14 days but can be as short as 3 days or as long as 30 days. Some infected persons have no recognized illness in that their infection cannot be ascertained by a blood test. Others may manifest only nonspecific symptoms (e.g., fever, headache, fatigue, and myalgia) that preclude a definitive diagnosis.

- If untreated or inadequately treated within weeks to months, the infection can progress from a "localized" or "early-disseminated" disease to a persistent disease (also called "late-disseminated" disease) that is manifested by intermittent swelling and pain of one or some joints (usually, the large, weight-bearing joints such as the knee). Accompanying diseases may be chronic axonal polyneuropathy, or encephalopathy accompanied by cognitive disorders such as sleep disturbance, fatigue, and personality changes. LD morbidity can be severe, chronic, and disabling. An ill-defined "post-Lyme disease syndrome" occurs in some persons after treatment for Lyme disease as I will discuss more at length in Chapter 3. Fortunately, LD is rarely, if ever, fatal.

How is LD transmitted?

From the site of inoculation, LD disseminates along three routes: cutaneous, lymphatic, and blood-borne. Early dissemination of the infection usually occurs days to weeks after the appearance of a solitary or secondary EM lesions. It can also manifest as a disease of the musculoskeletal system, the nervous system, or the heart. Musculoskeletal manifestations can include migratory pains of the muscles or the joints with or without joint swelling. Early neurologic manifestations include lymphocytic meningitis; cranial neuropathy, especially facial nerve palsy; and radiculoneuritis. Cardiac manifestations are rare but can include myocarditis and transient atrioventricular block of varying degree. I will later describe and illustrate most of these manifestations. However, for now, it will suffice to remember that there may be these three possible manifestations: muscular, neurologic, and cardiologic).

Ticks of the United States and their characteristics

Before dwelling more at length on the culprit for LD, it will be helpful to summarily review the various ticks encountered in the U.S. and their associated diseases. These are summarized in Table 2.1, showing the tick common name, the bacterium, the vector carrying it, and the diseases transmitted, where found, and what are the high-risk seasons.

As indicated earlier, Table 2.1 shows that LD can be caused in three separate ways by:

- *Borrelia burgdorferi* carried by the black-legged tick *Ixodes scapularis*;

- *Borrelia burgdorferi* carried by the western black-legged tick *Ixodes pacificus*; and

- *Borrelia mayonii* also carried by the blacked-legged tick *Ixodes scapularis*.

Note that *Borrelia burgdorferi* can be transmitted by five species of *Ixodes* ticks within the *Ixodes* family. In North America, they are: *Ixodes scapularis, Ixodes pacificus,* and *Ixodes cookei*. In Europe and Asia, they are *Ixodes ricinus* and *Ixodes persulcatus*, respectively. To limit the scope of this book, I have not discussed the latter two types of ticks.

When discussing LD, it is helpful to identify which tick transmits which pathogen causing the disease. Often, the variety (1) above is the one generally discussed. However, varieties (2) and (3) also deserve consideration, particularly since they recognize two different ticks and two different pathogens.

Lyme disease

Also, the same tick can carry several pathogens at the same time and cause several possible co-infections. For example, the black-legged tick *Ixodes scapularis* can transmit no less than seven pathogens and cause no less than six different diseases! Likewise, the western black-legged tick *Ixodes pacificus* can carry three different pathogens and cause three other infectious diseases, two of which (anaplasmosis, Lyme disease) being the same as for the tick *Ixodes scapularis*.

The multiplicity of pathogens transmitted by the several ticks further shows the number of other possible types of co-infectious diseases. While geography and season may help in limiting the number of such possibilities, the task of reaching an accurate diagnosis for any given tick-borne pathogen remains daunting.

Table 2.1 – Ticks of the United States and their characteristics

Tick common name	Vector	Bacterium	Disease(s) caused	Where found in the U.S.?	High-risk seasons
Black-legged	*Ixodes scapularis*	o *Borrelia burgdorferi* o *Borrelia mayonii* o *Anaplasma phagocytophilum* **o *Borrelia miyamotoi*** o *Ehrlichia chaffensis, Ehrlichia ewingii, Ehrlichia muris eauclairensis* o *Babesia microti* o *Powassan* virus	o *Lyme disease* o *Lyme disease* (newly discovered form) o Anaplasmosis o A form of relapsing fever o *Ehrlichiosis* o *Babesiosis* o Powassan virus disease	East, upper Midwest, and mid-Atlantic	Spring, Summer, Fall (any time when temperatures are above freezing). All tick life stages bite humans, but lymphs and adults are most commonly found on people
Western black-legged	*Ixodes pacificus*	o *Anaplasma phagocytophilum* o *Borrelia burgdorferi* o *Borrelia miyamotoi* (very likely)	o *Anaplasmosis* o *Lyme disease* o *Borrelia miyamotoi* disease (a form of relapsing fever)	Pacific Coast States	Larvae and nymphs often feed on lizards, birds, and rodents. Adults feed on deers. All life stages bite humans, but lymphs and adult females are most often reported
American dog	***Dermacentor variabilis***	o *Francisella tularensis* o *Rickettsia rickettsii*	o Rocky mountain spotted fever	East of the Rocky Mountains. Also, limited areas of the Pacific Coast	Bites most likely by adult females

Lyme disease

Brown dog	*Ripicephalus sanguineous*	*Rickettsia rickettsii*	Spotted mountain fever	Southwestern and Western States along the border with Mexico	May bite humans and other mammals
Ground hog (aka Woodchuck)	*Ixodes cookei*	Powassan virus	Powassan virus disease	Throughout Eastern half of U.S. States	Feed on a variety of warm-blooded animals, including groundhogs, skunks, squirrels, raccoons, foxes, weasels, and occasionally on people and domestic animals
Golf coast	*Amblyomma maculatum*	*Rickettsia parkeri*	*Rickettsia parkeri rickettsiosis* (a form of spotted fever)	Southeastern, mid-Atlantic States, and southern Arizona	Larvae and nymphs feed on birds and small rodents. Adults feed on deers and other wildlife
Lone star (Very aggressive. Adult female distinguished by white dot or "lone star" on her back.) Note: Allergic reactions may be associated with consumption of red (mammalian) meat.	*Amphylomma americanum*	o *Ehrlichia chaffensis* o *Ehrlichia ewingii* o *Francisella tularensis* o Heartland virus o Bourbon virus o Southern tick	o Ehrlichiosis o Ehrlichiosis o Tularemia o Heartland virus disease o Bourbon virus disease o Southern tick-associated rash illness (STARI)	East, more common in the South	Early Spring through late Fall. Lymphs and adult females most frequently bite humans
Rocky mountain wood	*Dermacentor andersoni*	o *Rickettsia rickettsii* o Colorado tick fever virus o *Francisella tularensis*	o Rocky Mountain spotted fever o Colorado tick fever disease o Tularemia	Rocky mountain States	Larvae and nymphs feed on small rodents. Adults feed primarily on rodents and are primarily associated with pathogen transmission to humans
Soft	***Ornithodoros Spp***	o *Borrelia hermsii* o *Borrelia turicatae*	Tick-borne relapsing fever (TBRF)	Throughout the western half, rustic cabins, cave exposure	Emerge at night and feed briefly while people are sleeping

Source: Adapted from CDC&P (2018).

Really,... what is LD?

Strictly speaking, LD is the disease transmitted to humans by the bite of infected black-legged ticks *Ixodes scapularis* (Figures 2.4 and 2.5), which transmits the bacterium *Borrelia burgdorferi* (less frequently, *Borrelia mayonii*). It can also be caused by *Borrelia burgdorferi* carried out by the western black-legged tick *Ixodes pacificus*.

Figure 2.4 – The adult deer tick *Ixodes scapularis*, the primary vector of Lyme disease in Central and Eastern U.S. but also elsewhere such as in Europe

Source: (U.S.) Centers for Disease Control & Prevention (CDC&P), Public Health Image Library PHIL #6631

Figure 2.5 – Adult deer tick, *Ixodes scapularis*

According to some scientists, some individuals with *Borrelia burgdorferi* can remain seropositive (that is, still have the bacteria in their blood) for decades, even after successful treatment with antibiotics and resolution of their clinical symptoms. By contrast, others believe that this is the normal humoral immune response and not a continuation of active chronic infection, especially if all key symptoms remain at bay. I will have more to say on this topic in the following chapters, beginning with Chapter 3 on the signs and symptoms of LD.

The developmental stages of *Ixodes scapularis*

The developmental stages of *Ixodes scapularis* are illustrated in Figures 2.6. and 2.7. Beginning from the left and progressing to the right, these stages include the unfed larva, engorged larva, unfed nymph, engorged nymph, unfed male and female, and lastly the partially engorged female (at the bottom right).

The nymphal stages are those of importance at the beginning of LD. This is further illustrated in Figure 2.6 with an indication of size (relative to a U.S. 25 cent coin).

Figure 2.6 – Developmental stages of *Ixodes scapularis*

The life cycle of the deer tick *Ixodes scapularis*

The lifecycle of the *Ixodes scapularis* tick is diagrammed in Figure 2.8 following the seasons. The egg is laid down in Spring and, after one month, transforms itself into a larva, which feeds once every two days preferably on a mouse host from Summer to Fall. It then hibernates through the Winter season till the next Spring when it transforms itself into a nymph. Until the next Fall, the nymph feeds itself 3-4 times a day, preferably on a mouse host.

Sexual distinction then occurs between Fall and Winter when the nymphs become adults and mating takes place. Feeding is now once a day with a preferred shift from a mouse to a deer host. Following hibernation, and three weeks into Spring, eggs are deposited and the adults die. The biannual cycle then repeats itself.

On the other hand, the *Ixodes persulcatus* tick which spreads the *Borrelia afzelii* and *Borrelia garinii* spirochetes is thought more likely to bite humans in its adult stage, with the nymphs seemingly reluctant to bite humans. The larva is usually infected (<5%) and is unlikely to either feed on humans or transmit infection to humans.

Figure 2.7 – Further illustration of *Ixodes scapularis* developmental stages

A new view on LD

A new view on LD has recently been advanced in which rodents (not deers) hold the key to the annual risk of contracting LD. The ecological determinants of LD over a 13-year period in southeastern New York, a hot zone for the disease, were examined. Combining field data with computer simulations, trends in inter-annual variation were analyzed, yielding two powerful predictors of entomological risk of LD in any given year: (1) the abundance of tick hosts (white-footed mice and chipmunks) in the previous year and (2) the abundance of acorns (which sustain the rodents) two years out. These findings have upset the long-held view that deer and climate are the best indicators of disease risk.

In this view, *Ixodes scapularis* larvae hatch in midsummer, and acquire infection after feeding on an infected mouse or other small animal. Larvae detach after several days of feeding, morph into nymphs, and enter a nearly year-long dormant stage. After another round of feeding, nymphs fall off and molt into adults, which prefer the blood of larger mammals. While larvae and nymphs can acquire and transmit infection, people are most likely to contract LD from nymphs.

From 1991 to 2004, the researchers collected temperature and precipitation data, and estimated the abundance of acorns and animals on six plots of land. From this 13-year data set, they developed computer models to estimate how each of the 11 variables (including multiple climate and deer indexes) contributed to yearly variations in the density of infected ticks and, thus, the risk of human exposure.

Figure 2.8 – Life cycle diagram of the deer tick *Ixodes scapularis*

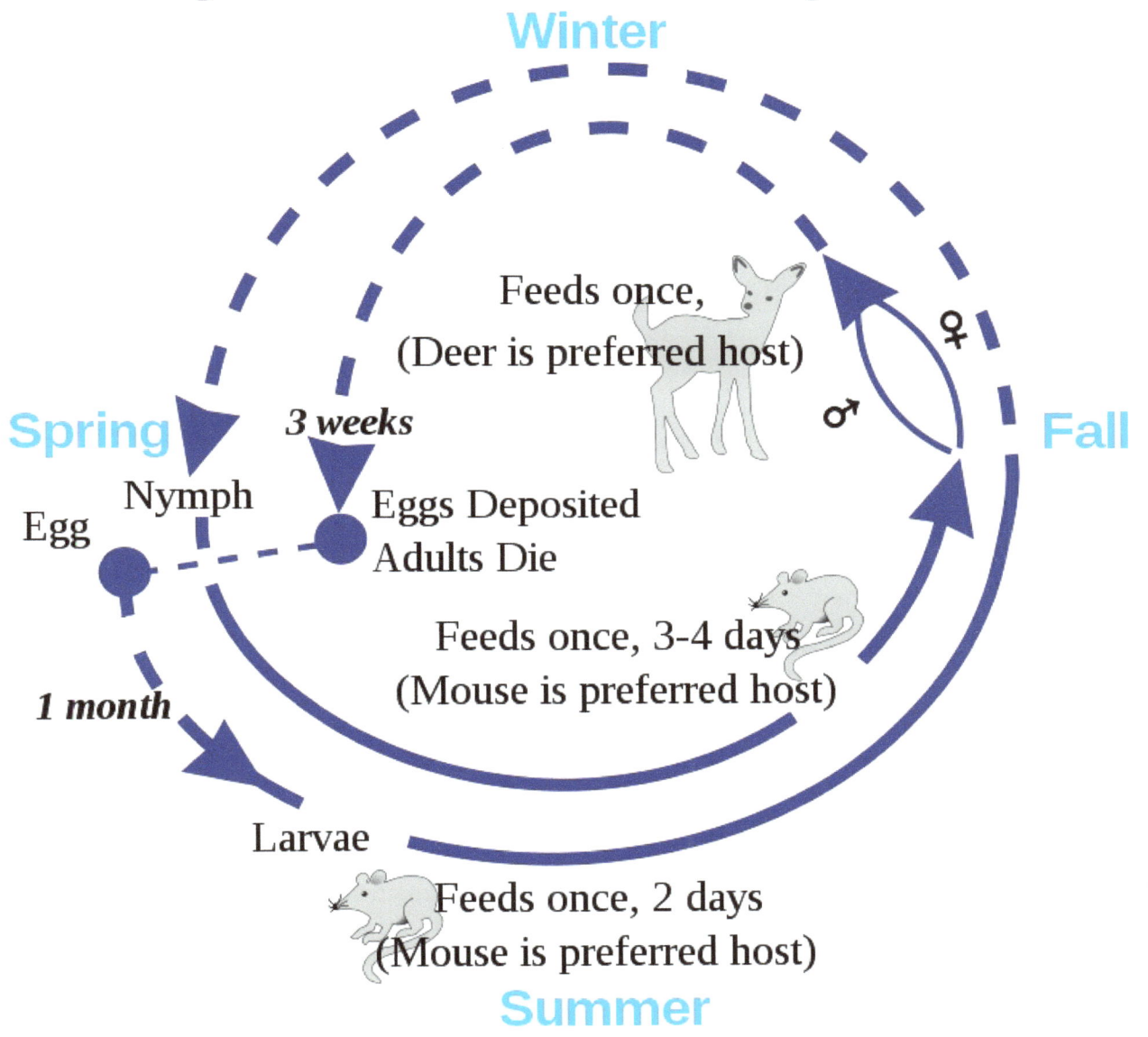

Source: Philg88

While none of the climate variables influenced nymphal infection prevalence, higher temperatures in the previous year and precipitation patterns in the current year had weak, though unexpected, effects on total density and density of infected nymphs. It is thought that higher temperatures keep tick populations down, but the models showed them increasing both total density and density of infected nymphs.

Figure 2.9 – Circulation of pathogenic European genospecies of Borrelia *burgdorferi sensu lato*

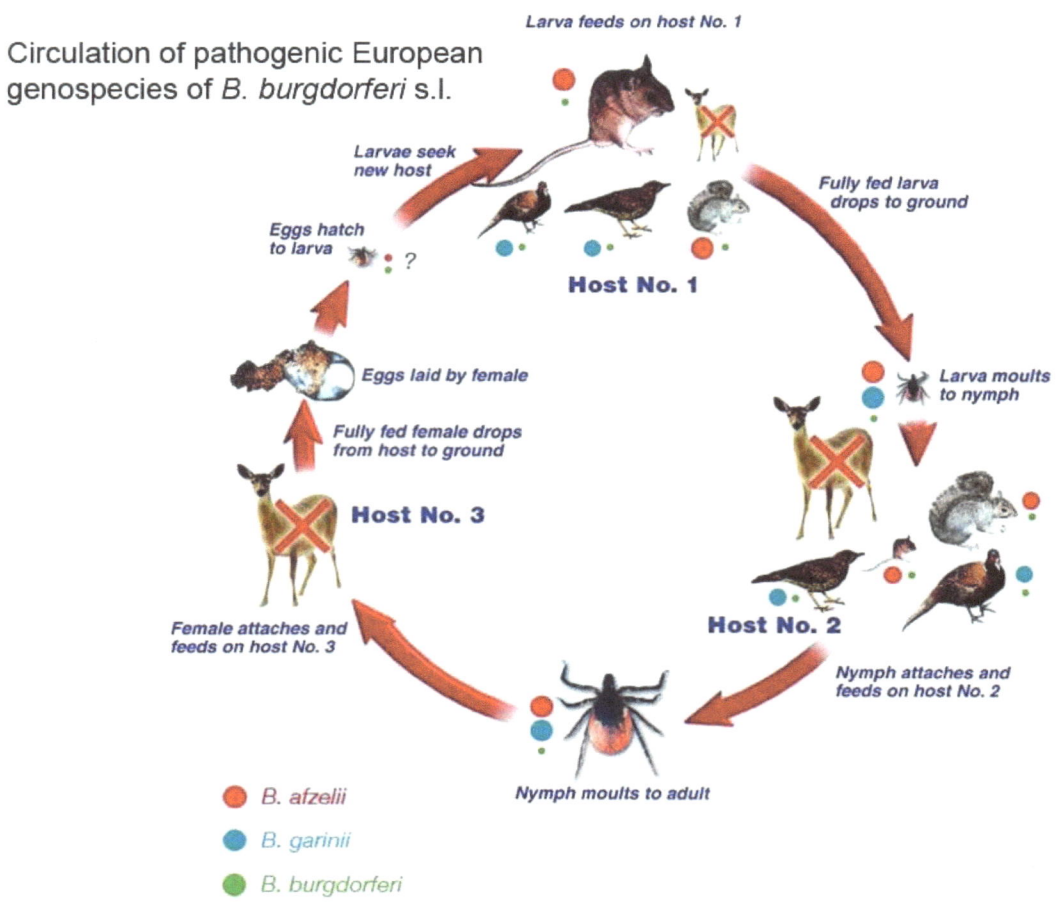

Although tick survival is expected to rise with precipitation, the models found the highest tick numbers at intermediate precipitation levels. These inconsistencies can be explored by incorporating other variables with documented effects into the approach outlined here. Also, surprisingly, the researchers found that even a 3-fold variation in deer numbers had no impact on subsequent nymph abundance.

Density of infected nymphs—the principal determinant of LD risk—varied significantly from year to year, fueled mostly by large fluctuations in total nymph density that, in turn, depended mostly on fluctuations in abundance of acorns, mice, and chipmunks. Interestingly, though chipmunk densities are generally lower than mice, their numbers were the best predictor of total nymph density in the subsequent year, likely reflecting their inferior grooming skills. Overall, the results found that acorns were the best predictor of LD risk—stemming from their crucial role in supporting white-footed mice, chipmunks, and likely other small animals, which provide large reservoirs for *Borrelia burgdorferi*. Acorns will not be a universal predictor of risk, the researchers acknowledge, since the disease can occur in areas without oaks. But the strength of these findings suggests that the observed link between increased LD risk and high rodent densities indicates that important food sources—or predators—of the rodent hosts of nymphs will be valuable predictors of disease risk.

Notwithstanding the above observations and findings, symptoms remain the really most important factor, which I will discuss at length in Chapter 4.

Can LD be inherited?

No! LD cannot be inherited. However, the risk of certain complications of the condition may be influenced by inherited genetic factors, but the inheritance pattern is unknown.

The terrible invader, evader... and great imitator

Borrelia is a terrible invader and evader in addition to being a great imitator:

- *Invader:* *Borrelia* is capable of burrowing and spreading throughout most of the body's tissues, including the blood-brain barrier (BBB) – that highly selective, semi-permeable border that separates the circulating blood from the brain and extracellular fluid in the central nervous system (CNS). The BBB allows the passage of some molecules such as glucose, water, and amino acids that are crucial to neural function but antibiotics cannot easily cross it. However, it allows *Borrelia* and other infections like *Babesia* to attack the brain and the CNS unimpeded. *Borrelia* implants itself in tissues and even within individual cells to avoid being consumed and lysed by the immune system. When it does find a place to settle down and multiply, the microbe secretes a protein mesh called a "biofilm" which shields it and its progeny from direct exposure to many antibiotics and other compounds that might otherwise harm the infection.

- *Evader:* The microbe is also adept at countering the adaptive immune system. Through several mechanisms, including a constant variation in its surface proteins (or antigens), it can distract and evade the immune system. Usually, by the time the body can create antibodies to target and neutralize *Borrelia*, its surface has already changed so much that the new antibodies are no longer effective. In serum. *Borrelia* has the ability to inactivate C3, a protein marker which would normally trigger the immune response for free-floating microbes.

- *Imitator:* LD is one of the most misunderstood and widely growing illness in the country and abroad. Because its symptoms resemble those of so many other diseases (and because the medical community remains misinformed or incompletely informed), it has been recognized as "the great imitator". It can mimic the symptoms of fibromyalgia, chronic fatigue syndrome (CFS), multiple sclerosis (MS), amyotrophic lateral sclerosis (ALS) or Lou Gehring disease, Parkinson's disease (PD), Alzheimer's disease (AD), as well as more than some 350 other diseases. More on this in later Chapters.

LD has also generally been misinterpreted. As a consequence, stricken patients are not diagnosed correctly and may not receive the treatment necessary to restore their health.

Sidebar 2.1 provides a brief look at tick-borne diseases of the United States. My aim there is showing you the multiplicity of such diseases and the associated difficulties in reaching the correct diagnosis.

Take-home points

- Lyme disease, also known as Lyme *borreliosis*, is an infectious disease caused by the *Borrelia bacterium*. It was originally mistaken for juvenile rheumatoid arthritis. The bacterium involved, *Borrelia burgdorferi,* was first described in 1981. There are three varieties depending on the pathogen transmitted and the carrier tick.

- Lyme is the most common disease spread by ticks in the Northern Hemisphere. Annually, it affects about 300,000 people in the U.S. and 85,000 people in Europe. Infections are most common in the Spring and early Summer.

- Even after successful treatment with antibiotics and resolution of their clinical symptoms, some patients can still have the bacteria in their blood for decades. However, when the key symptoms are kept at bay, it is also believed that this is a normal humoral immune response and not a continuation of active chronic infection.

- Through a number of mechanisms, the microbe is adept at countering, distracting, and evading the adaptive immune system. It is also capable of crossing the blood-brain barrier (that gate that separates the circulating blood from the brain) and infect the brain.

- Lyme cannot be inherited. However, the risk of certain complications of the condition may be influenced by inherited genetic factors, but the inheritance pattern is unknown.

- While antibiotics cannot easily pass the blood-brain barrier, *Borrelia* and other infections can cross it and attack the brain and the central nervous system.

- When *Borrelia* settles down and multiplies, it secretes a protein mesh (a "biofilm") which shields it and its progeny from direct exposure to many antibiotics and other compounds.

- A new view on LD has recently been advanced in which rodents (*not* deers) hold the key to annual risk. Two powerful predictors of entomological risk of LD in any given year were found to be: (1) abundance of tick hosts (white-footed mice and chipmunks) in the previous year and (2) abundance of acorns (which sustain the rodents) two years out. These findings upset the long-held view that deer and climate are the best indicators of disease risk.

- There are various tick-borne diseases encountered across the U.S. Their causal bacteria and the corresponding vector ticks have been summarized in Table 2.1. If you are locating (or

Lyme disease

relocating) somewhere in the U.S., it behooves you to become acquainted with the specific tick-borne diseases that prevail in that State (or region).

Sidebar 2.1
A brief look at tick-borne diseases of the United States

In the U.S., several ticks carry pathogens that can cause human diseases including:

Anaplasmosis: This disease is transmitted to humans by tick bites primarily from the black-legged tick (*Ixodes scapularis*) in the northeastern and upper midwestern U.S. and the western black-legged tick (*Ixodes pacificus*) along the Pacific coast.

Babesiosis: This disease is caused by microscopic parasites that infect red blood cells. Most human cases of *babesiosis* in the U.S. are caused by *Babesia microti,* which is transmitted by the black-legged tick (*Ixodes scapularis*) and is found primarily in the Northeast and upper Midwest.

Borrelia miyamotoi: The bacterium *Borrelia miyamotoi* has recently been described as a cause of illness in the U.S. It is transmitted by the black-legged tick (*Ixodes scapularis*) and has a range similar to that of Lyme disease.

Bourbon virus: disease: The virus has been identified in a limited number of patients in the Midwest and Southern U.S. At this time, we do not know if the virus might be found in other areas of the United States.

Colorado tick fever: This disease is caused by a virus transmitted by the Rocky Mountain wood tick (*Dermacentor andersoni).* It occurs in the Rocky Mountain states at elevations of 4,000 to 10,500 feet.

Ehrlichiosis: This disease is transmitted to humans by the lone star tick (*Ambylomma americanum*). It is found primarily in the south-central and eastern U.S.

Heartland virus disease: Cases of the disease caused by this virus have been identified in the Midwestern and Southern U.S. Studies suggest that Lone Star ticks can transmit the virus. It is unknown if the virus may be found in other areas of the U.S.

Lyme disease: The bacterium *Borrelia mayonii* has recently been described as a cause of illness transmitted by black-legged ticks (*Ixodes scapularis*). *Borrelia mayonii* is a new species that is the only species besides *Borrelia burgdorferi* known to cause Lyme disease in North America. This disease is transmitted by the black-legged tick (*Ixodes scapularis*) in the Northeastern U.S. and upper Midwestern U.S. and the Western black-legged tick (*Ixodes pacificus*) along the Pacific coast.

Powassan disease: This disease is transmitted by the black-legged tick (*Ixodes scapularis*) and the groundhog tick (*Ixodes cookei*). Cases have been reported primarily from Northeastern states and the Great Lakes region.

Rickettsia parkeri rickettsiosis: This disease is transmitted to humans by the Gulf Coast tick

Lyme disease

(*Amblyomma maculatum*).

Rocky Mountain spotted fever (RMSF): This disease is transmitted by the American dog tick (*Dermacentor variabilis*), Rocky Mountain wood tick (*Dermacentor andersoni*), and the brown dog tick (*Rhipicephalus sanguinineus*) in the U.S. The brown dog tick and other tick species are associated with RMSF in Central and South America.

Southern tick-associated rash illness (STARI): This disease is transmitted via bites from the lone star tick (*Ambylomma americanum*) found in the Southeastern and Eastern U.S.

Tick-borne relapsing fever (TBRF): It is transmitted to humans through the bite of infected soft ticks. TBRF has been reported in 15 states: Arizona, California, Colorado, Idaho, Kansas, Montana, Nevada, New Mexico, Ohio, Oklahoma, Oregon, Texas, Utah, Washington, and Wyoming) and is associated with sleeping in rustic cabins and vacation homes.

Tularemia: This disease is transmitted to humans by the dog tick (*Dermacentor variabilis*), the wood tick (*Dermacentor andersoni*), and the lone star tick (*Amblyomma americanum*). *Tularemia* occurs throughout the U.S.

364D rickettsiosis (alternative proposed name: *Rickettsia phillipi*): This disease is transmitted to humans by the Pacific Coast tick (*Dermacentor occidentalis*). This is a new disease that has been found in California.

Table 2.2 summarizes the several ticks causing the illnesses listed above. In particular, it is seen that LD bacteria are *Borrelia burgdorferi* and *Borrelia mayonii* that are both transmitted through the *Ixodes scapularis* tick. That disease can also be transmitted by the *Ixodes pacificus* tick.

Table 2.2 – Tick-associated diseases of the United States

Vector	Bacterium	Disease(s) transmitted
Ambyloma americanum (lone star tick)	o Heartland virus	o *Ehrlichiosis* o *Rickettsia parkeri rickettsiosis* o Southern Tick-Associated Rash Illness (STARI) o *Tularemia*
Ambyloma maculatum		o *Rickettsia parkeri rickettsiosis*
Dermacentor andersoni (wood tick)		o Colorado tick fever o Rocky Mountain spotted fever o *Tularemia*
Dermacentor occidentalis		o *364D rickettsiosis* (aka *Rickettsia philippi*)
Dermacentor variabilis (dog tick)		o Rocky Mountain spotted fever o *Tularemia*
Ixodes cookei (groundhog tick)		o Powassan disease

Lyme disease

Ixodes pacificus (black-legged tick)		o *Anaplasmosis* o **Lyme disease**
Ixodes scapularis (black-legged tick)	o *Borrelia burgdorferi* o *Borrelia mayonii* o *Borrelia miyamotoi* o *Babesia microti* X	o **Lyme disease** o **Lyme disease** o *Anaplasmosis* o *Babesiosis* o Powassan disease
Lone star tick	o Heartland virus	o *Ehrlichiosis*
Rhipicephalus sanguineus		o Rocky Mountain spotted fever
Soft tick		o Tick-borne relapsing fever (TBRF)

Source: Adapted and augmented from "Tick-borne Diseases of the United States: A Reference Manual for Health Care Providers", Fifth Edition (2018).

An abridged geographical risk for the above diseases across the U.S. is also shown below:

Eastern U.S.: *Ehrlichiosis,* Southern tick-associated rash illness (STARI).

Gold coast: *Rickettsia parkeri rickettsiosis*.

Great Lakes region: Powassan disease.

Individual States (Arizona, California, Colorado, Idaho, Kansas, Montana, Nevada, New Mexico, Ohio, Oklahoma, Oregon, Texas, Utah, Washington, and Wyoming): Tick-borne relapsing fever (TBRF).

Midwestern: Bourbon virus, heartland virus.

Northeastern U.S.: *Anaplasmosis, babesia,* Lyme disease.

Pacific coast U.S.: *Anaplasmosis, 364D rickettsiosis* (aka *Rickettsia phillipi*).

Rocky Mountain States (at elevations of 4,000 to 10,500 feet): Colorado tick fever, Rocky Mountains spotted fever (RMSF). Note: RMSF is also present in Central and South America.

South Central U.S.: *Ehrlichiosis*.

Southern U.S.: Bourbon virus, heartland virus, Southern tick-associated rash illness (STARI).

Throughout the U.S.: *Tularemia*.

Upper midwestern U.S.: *Anaplasmosis, Babesia,* Lyme disease.

3

What are CLD, CLDC, ... and PLDS?

Recall that Lyme disease is caused by *Borrelia burgdorferi* and other closely related species of bacteria. The infection is commonly contracted through a tick sting (although most people think of it as a bite, but technically it is a sting), which may not even be noticed. Within the first few days to weeks of the infection, a rash called *erythema migrans* (EM) may develop. It is generally red and circular but, as it subsides, it may present a "bull's-eye" pattern. Although the rash is a major sign of acute Lyme disease, many infected people do not get it, that is, while the rash is a definite indication for LD, its absence does not mean the absence of LD. During this early stage, it is referred to as "localized" because it has not yet spread to other areas of the body and seems like the flu (aches, pains, and swollen glands). Like other short-term infections, it may last a few weeks. It is easier to diagnose than treat. The recent history of the patient's exposure and a total body skin exam for rash are followed by a round of antibiotics after which the affected individual is usually (but not generally) well again within 4-6 weeks.

What is chronic Lyme disease?

Chronic Lyme disease (CLD), also called "late" Lyme, is much different than the localized Lyme disease I alluded to above. The term "chronic Lyme disease" has been used to describe people with different illnesses. While the term is sometimes used to describe illness in patients with LD, it has also been used to describe symptoms in people who have no clinical or diagnostic evidence of a current or past infection with *Borrelia burgdorferi*. Because of the confusion in how the term CLD is employed, and the lack of a clearly defined clinical definition, many experts in this field do not support its use.

Based on the combined expertise of its membership and its systematic review of 250 peer-reviewed papers in the international scientific literature, the International Lyme and Associated Diseases Society (ILADS) issued on 16 December 2019, a concept paper. That publication provided its "evidence-based" definition of CLD as a *"multi-system illness with a wide range of symptoms and/or signs that are either continuously or intermittently present for a minimum of six months. The illness is the result of an active and ongoing infection by any of several pathogenic members of the Borrelia burgdorferi sensu lato complex (Bbsl). The infection has variable latency periods and signs and symptoms may wax, wane, and migrate"*. ILAD further subdivided CLD into two sub-categories (a) "untreated" CLD (CLD-U) and previously treated CLD (CLD-PT), the latter requiring that *"CLD manifestations persist or recur following treatment and are present continuously or in a relapsing/remitting pattern for a duration of six months or more*". It is hoped that use of this CLD definition, if accepted, will promote a better understanding and facilitate future research of this infection.

However,... CLD is not an official diagnosis!

The term "chronic Lyme disease" actually is not an official diagnosis. It categorizes patients who suffer from a wide variety of vector-borne diseases together with the primary and secondary infections. The most common co-infections include: *anaplasmosis, babesiosis, bartonellosis, Borreliosis,* Colorado tick fever, *ehrlichiosis* (my acronym **AB3CE**), and perhaps many others. Beyond AB3CE, other common primary co-infections that have been linked with CLD include chlamydia pneumonia, coltivirus disease, mycoplasma pneumonia, Powassan virus disease, Q-Fever (my acronym **AC2MPQ**), and perhaps others. CLD more accurately represents the wide range of infections patients can experience. It better illustrates the complexity and uniqueness of the condition burdening every single patient. Often, these patients will have a resistant *Borrelia* infection that is almost always followed by multiple other infections.

In CLD, signs of illness may appear gradually over time or may have never entirely subsided with earlier treatment. In some cases, major physical or emotional stress brings the disease to the forefront. Problems like joint pain, horrible fatigue, and trouble thinking may develop and become more debilitating over time.

Blood tests often miss this type of Lyme since the acute phase is past. Further, the infection can impair or even suppress the immune system so the body does not create enough antibodies to fight the infecting bacteria. The real source of poor health can remain a mystery, yet, the symptoms may be right there. When *Borrelia burgdorferi* is in the system for a while, it burrows its way into organs and tissues. This can give degenerative health issues that look like something else.

What are the signs and symptoms of CLD?

The four most common symptoms are (my acronym **(FJAB)**:

- **Debilitating chronic fatigue** (also called **chronic fatigue syndrome, CFS**): Its cause is not known. It is due in part to the taxing of the immune system because the bacteria trigger the immune cells to produce chemicals called "cytokines" that generate inflammation to fight the infection. CLD is a last-resort diagnosis when a reason for it cannot be found. Though it may feel reassuring to get a diagnosis, the wrong one can stand in the way of getting better. Nonetheless, for many (not all) people, the root cause of CFS could be undetected CLD (for additional details on CFS, see Sidebar 3.1).

- **Joint pain:** Irrespective of a previous healthy condition and active way of life, Lyme damages the joints accompanied by crippling pain. The bacteria colonize the joints and connective tissues (cartilage and ligaments), bind to collagen in these tissues, and break it down, destroying it as they multiply. Further, they interfere with the body's efforts to regenerate these tissues. Note that the bacteria are completely dependent on the individual for nutrients. Since they cannot make certain proteins, they siphon them from the joints leaving the ligaments brittle and the joints stiff and inflamed. It appears (but is not) arthritis (Sidebar 3.2).

- **Autoimmune diseases:** The immune system attacks the tissues and the bacteria do whatever they can to protect themselves from it. They have learned to hide in the host's cells. This happens because some of the proteins in the bacteria look like the host's own. The immune system then starts to view healthy cells as foreign invaders and it attacks them thinking it gets rid of the Lyme. In some individuals, the problem can escalate to rheumatoid arthritis in which the body attempts to stomp out joint cells where the bacteria live. Separately, the tails of the bacteria look like the cells of the host's myelin sheath (these are the cells that cover and insulate your nerves). The bacteria can trick the immune system into attacking the healthy myelin cells and even block the body's ability to regenerate the myelin. Without this protective nerve covering, inflammation takes place with intense pain, a process that looks exactly like, but is not, multiple sclerosis (MS) (Sidebar 3.3).

- **Brain fog and cognitive issues:** The bacteria are able to cross the protective blood-brain barrier and create new problems that may mimic other diseases. After getting into the brain, they inhabit glial cells and neurons, and render them dysfunctional, leading to anxiety and depression. They may also interfere with the host's neurotransmitters or nerve messengers, disrupting the production and use of dopamine, and causing what can look like (but is not!) Parkinson's disease (PD). Further, the microbes can increase the production of another neurotransmitter (called octopamine) that can accumulate in the brain, raise blood pressure, and cause headaches. Lastly, when this insidious microorganism is killed, it sheds parts of itself as endotoxins, which may interfere with the brain's chemistry. A buildup of toxins can give intense brain fog and memory issues (Sidebar 3.4).

The potential for co-infections varies by location as well as exposure to various species of ticks. Since there is a potential for patients to experience several tick bites, including those of different species, additional microorganisms are also commonly transmitted via tick bites. Many of these other infections complicate treatment and increase the risk for auto-immune diseases, depending on the patient's genetic background. To repeat, patients with CLD never have just the one infection by *Borrelia burgdorferi*. They also have multiple co-infections that have gone untested and incompletely or totally untreated for extended periods of time. Like for viral, fungal, and parasitic infections, antibiotics are not effective for many of these infections that entered the body in the initial tick bite.

What is CLDC?

Chronic Lyme disease complex (CLDC) is not just CLD. It involves the spirochete which causes Lyme, *Borrelia burgdorferi*, and also associated co-infections, which enter the body with it in the initial tick bite (my earlier acronyms AB3CE and C2MPQ). CLDC can also include secondary infections that occur as a result of a compromised immune state. This is important to keep in mind because antibiotics do not work against viral infections and, similarly, fungal infections require anti-fungal medication, and so on. The secondary and co-infections must be identified in order for treatment to be effective. This is the number one reason why antibiotics alone cannot be used to treat these patients.

Further, as indicated in the sub-title of this book and in Chapter 2, *Borrelia* has a number of mechanisms for invading, evading, and distracting the immune system. It is capable of burrowing and spreading throughout most of the body's tissues, including the blood-brain barrier (BBB). Antibiotics cannot easily pass the BBB, but *Borrelia* and other infections attack the brain and central nervous system unimpeded. It implants itself in tissues and within individual cells to avoid being consumed and lysed by the immune system. The microbe is also adept at countering the adaptive immune system through constant variation in its surface proteins, or antigens, a feat enabled by its complex single chromosome and its circular DNA plasmids. Usually, by the time the body can create antibodies to target *Borrelia*, its surface has already changed so much that the new antibodies are no longer effective. In serum, *Borrelia* has the ability to inactivate C3, a protein marker which would normally trigger the immune response for free-floating microbes. When *Borrelia* does find a place to settle down and multiply, the microbe secretes a protein mesh called "biofilm" which shields it and its progeny from direct exposure to many antibiotics and other compounds that might otherwise harm the infection.

Thus, while the spirochete *Borrelia* is responsible for causing primary LD, it is usually not the only factor to consider when approaching treatment for CLDC. When a tick bites, various bacteria and parasites enter the wound besides just *Borrelia*. These simultaneous infections, called "co-infections", can include *Babesia*, Human herpes virus (HHV)-6, Epstein-Barr virus (EBV), and Rocky Mountain Spotted Fever (RMSF), to name just a few. They are able to gain a firmer foothold in the body and potentially become chronic since *Borrelia* excels at distracting and depressing the host immune system. "Secondary infections" can also be picked up after the initial bite due to a depressed immune response from the infected person. The result is a condition that is chronic and can linger for months or even years, typically worsening over time without proper treatment. The simultaneous presence of multiple different infections in the body seriously complicates any potential treatment, increases the risk for autoimmune diseases depending on the patient's genetic background, and can also be a precursor to various cancers. Treating these infections in their totality is essential to full recovery and for the patient to achieve health, which is why often antibiotics alone do not work clinically.

Again, to repeat, patients with CLDC never have just the one infection by the spirochete *Borrelia burgdorferi*. They also have multiple co-infections and secondary infections that have gone untested or/and insufficiently treated or even untreated for extended periods of time. Like for viral, fungal, and parasitic infections, antibiotics are not effective for many of these infections that entered the body in the initial tick bite. What needs to be done is the subject of the following Chapters.

... and what is PLDS?

My purpose here is not to dwell on the Lyme alphabet soup, but it is important to distinguish between chronic Lyme disease (CLD) and post-treatment Lyme disease syndrome (PLDS). (PLDS was previously, though incorrectly, recognized as CLD, but the medical community no longer recognizes that diagnosis and it is now considered an outdated term.) Some healthcare providers claim that PLDS is caused by persistent infection, but this is not believed to be true because of the inability to detect infectious organisms after standard treatment. Rather, PLDS describes the continued ongoing resistance to antibiotics and continued symptoms. Further, PLDS does not define CLDC, which encompasses the breadth of infections present, regardless of whether standard-of-care treatment for Lyme *Borrelia* was given or not.

While CLD and PLDS may seem similar, they are very different from each other and knowing the difference may be essential for achieving symptom remission. I did not say cure …only symptom remission! PLDS is a diagnosis given to patients who continue to experience symptoms like depression, fatigue, and muscle/joint aches after being diagnosed and treated for LD with the recommended antibiotic regimen. In fact, the Lyme *borrelia* may not always be present in a CLD patient as the term refers to a wide range of tick-borne plus other secondary infections and complications that the patient may be dealing with. This infection load and the complications associated with it vary greatly from patient to patient. Like in most other areas of medicine wherein "one size does not fit all", a more personalized approach often leads to an improved patient outcome.

The problem with PLDS is that conventional medicine only recognizes Lyme *borrelia* as a causative factor, but provides no treatment for the primary and secondary infections and other complicating genomic risks. On top of that, PLDS only covers patients who have already been treated for LD. The PLDS diagnosis completely fails to address patients who are suffering from the many different infections seen clinically.

PLDS is always characterized by a Lyme *borreliosis* infection. While it may involve many other infections including viral, bacterial, and fungal, those infections are controversially not recognized as part of PLDS. These various infections are rarely tested properly and, while approximately 10% of those who are diagnosed and treated for LD with the standard antibiotic regimen will experience PLDS, there is no conventionally accepted treatment as the symptoms are expected to subside over time.

Dr. Burgdorfer was critical of the path LD research had taken over the past 30 years. He believed that *Borrelia burgdorferi* was a persistent infection, and that the current serological testing methodologies needed to be started anew without any knowledge of the results sought after.

Sidebars 3.1-3.4 will respectively expound on chronic fatigue syndrome, joints pain, autoimmune diseases, and brain fog.

Take-home points

- While *Borrelia* is responsible for causing primary Lyme disease, it is usually not the only factor. When one or several ticks perhaps from different species do bite, various bacteria and parasites also enter the wound causing other opportunistic infections. The overall disease caused by both the primary and secondary infections is referred to as chronic Lyme disease (or syndrome). The potential for co-infections varies by location as well as exposure to various species of ticks.

- Chronic Lyme disease, also called "late" Lyme, is much different than the localized Lyme disease. The term "chronic Lyme disease" has been used to describe people with different illnesses and also to describe symptoms in people who have no clinical or diagnostic evidence of a current or past infection with *Borrelia burgdorferi*. Because of the confusion in how the term chronic Lyme disease is employed, and the lack of a clearly defined clinical definition, many experts in this field do not support its use. By contrast, the International Lyme and Associated Diseases Association (ILADS) advocates for CLD and has provided corresponding diagnosis and therapeutics guidelines (see Chapter 23 for more details).

- The term "chronic Lyme disease" actually is *not* an official diagnosis. It categorizes patients who suffer from a wide variety of vector-borne diseases together with the primary and secondary infections.

- In chronic Lyme disease, signs of illness may appear gradually over time or may have never entirely subsided with earlier treatment. Blood tests often miss this type of Lyme since the acute phase is past. Further, the infection can impair or even suppress the immune system so the body does not create enough antibodies to fight the infecting bacteria.

- When *Borrelia burgdorferi* is in the system for a while, it burrows its way into organs and tissues creating degenerative health issues.

- Chronic Lyme disease can cause a significant burden on patients more so than Lyme disease alone. It can linger for months or even years, typically worsening over time in the absence of proper treatment and leading to debilitating symptoms.

- Many of these other infections complicate treatment and increase the risk for autoimmune diseases depending on the patient's genetic background. They can also be a precursor to various cancers and even neurodegenerative diseases. Treating these infections in their totality as quickly as possible is essential to full recovery and to achieve health.

- Like for viral, fungal, and parasitic infections, antibiotics are not effective for many of the infections that entered the body in the initial tick bite(s).

- The four most common symptoms of chronic Lyme disease are: debilitating fatigue, joint inflammation and pain, autoimmune diseases, and brain fog and cognitive problems (acronym **FJAB**).

- Chronic Lyme disease complex involves the spirochete which causes Lyme, *Borrelia burgdorferi*, and also associated co-infections, which entered the body with it in the initial tick bite . It can also include secondary infections which occur as a result of a compromised immune state. Antibiotics alone cannot be used to treat the condition.

- Post-treatment Lyme disease syndrome is purportedly caused by persistent infection, but this is not believed to be true because we are unable to detect infectious organisms after standard treatment. Rather, it describes the continued ongoing resistance to antibiotics and continued symptoms.

Sidebar 3.1
Chronic fatigue syndrome

What is CFS?

Chronic fatigue syndrome (CFS), also called myalgic encephalomyelitis/chronic fatigue syndrome (ME/CFS) or systemic exertion intolerance disease (SEID), is a serious, long-term illness that affects many body systems, preventing the affected individual from doing usual activities or even sometimes getting out of bed.

What causes CFS?

The cause of CFS is unknown. There may be two or more possible causes working together to trigger the illness. Researchers are looking into the following possible causes:

- **Infection:** About 1 out of 10 people who develop certain infections, such as Epstein-Barr virus (EBV) and Q-fever, go on to develop SEID/CFS. Other infections have also been studied, but no cause has yet been found;

- **Immune system changes:** SEID/CFS may be triggered by changes in the way a person's immune system responds to stress or illness;

- **Mental or physical stress:** Many people with SEID/CFS have been under serious mental or physical stress before becoming ill;

- **Energy production:** The way that cells within the body get energy is different in people with SEID/CFS than in people without the condition. It is unclear how this is linked to developing the illness; and

- **Genetics or environmental factors** – They may play a role in the development of SEID/CFS.

Who is at risk for CFS?

While anyone can get CFS, it is most common in people aged 40-60 years. The illness affects children, adolescents, and adults of all ages. Among adults, women are affected more often than adult men. Whites are diagnosed more than other ethnicities. But many people with SEID/CFS have not been diagnosed, particularly among minorities.

What are the symptoms of CFS?

CFS symptoms can include:

- Severe fatigue that does not improve with rest;

- Sleep problems;
- Post-exertional malaise (PEM), where symptoms get worse after any physical or mental activity;
- Problems with thinking and concentrating;
- Pain; and
- Dizziness.

CFS can be unpredictable as symptoms may come and go. They may change over time, at times improving, at other times getting worse.

How is CFS diagnosed?

CFS can be difficult to diagnose as there is no specific test and other illnesses can cause similar symptoms. Other diseases must be ruled out before making a diagnosis. A thorough medical examination should include:

- A personal and family's medical history;
- A review of current illness(es), including present symptoms (How often? How bad they are? How long they have lasted? And how they affect the patient's life?);
- A thorough physical and mental status exam; and
- Blood, urine or other tests.

What are the treatments for CFS?

While there is no cure or approved treatment, some of the symptoms can be managed. A treatment plan should be agreed on between the patient, his/her family, and the treating physician, figuring out which symptoms are the most problematic and addressing them first. For example, if sleep problems affect the most, good sleep habits must be developed. If those do not help, medicines should be taken or a sleep specialist consulted. Strategies such as learning new ways to manage activity can also be helpful avoiding the "push and crash" approach when feeling better, doing too much, and then getting worse again.

Since the process of developing a treatment plan and attending to self-care can be hard in CSF, it is important to have support from family members and friends. Also, beware that some treatments promoted as cures are unproven, often costly, and could be dangerous.

Sidebar 3.2
Joints pain

Joints pain can be caused by many types of injuries or conditions. It may be linked to arthritis, bursitis, and muscle pain. No matter what causes it, joints pain can be very bothersome.

What causes joints pain?

- Autoimmune diseases such as rheumatoid arthritis and lupus;
- Bursitis;
- Chondromalacia patellae;
- Crystals in the joint -- gout (especially found in the big toe) and CPPD arthritis (pseudogout);
- Infections caused by a virus;
- Injury, such as a fracture;
- Osteoarthritis;
- Osteomyelitis (bone infection);
- Septic arthritis (joint infection);
- Tendinitis; and
- Unusual exertion or overuse, including strains or sprains.

What are the signs of joint inflammation?

These include:

- Swelling;
- Warmth;
- Tenderness;
- Redness; and

- Pain with movement.

Sidebar 3.3
Autoimmune diseases

Immune system disorders cause abnormally low activity or over activity of the immune system. In cases of immune system over activity, the body attacks and damages its own tissues (autoimmune diseases). Immune deficiency diseases decrease the body's ability to fight invaders, causing vulnerability to infections.

In response to an unknown trigger, the immune system may begin producing antibodies that instead of fighting infections, attack the body's own tissues. Treatment for autoimmune diseases generally focuses on reducing immune system activity. Examples of autoimmune diseases include but are not all present in CLD :

- **Rheumatoid arthritis (RA):** The immune system produces antibodies that attach to the linings of joints. Immune system cells then attack the joints, causing inflammation, swelling, and pain. If untreated, rheumatoid arthritis gradually causes permanent joint damage. Treatments for rheumatoid arthritis can include various oral or injectable medications that reduce immune system over activity.

- **Systemic lupus erythematosus (SLE):** People with lupus develop autoimmune antibodies that can attach to tissues throughout the body. The joints, lungs, blood cells, nerves, and kidneys are commonly affected in lupus. Treatment often requires daily oral prednisone, a steroid that reduces immune system function.

- **Inflammatory bowel disease (IBD):** The immune system attacks the lining of the intestines, causing episodes of diarrhea, rectal bleeding, urgent bowel movements, abdominal pain, fever, and weight loss. Ulcerative colitis (UC) and Crohn's disease (CD) are the two major forms of IBD. Oral and injected immune-suppressing medicines can treat IBD.

- **Multiple sclerosis (MS):** The immune system attacks nerve cells, causing symptoms that can include pain, blindness, weakness, poor coordination, and muscle spasms. Various medicines that suppress the immune system can be used to treat MS.

- **Type 1 diabetes mellitus (T1DM):** Immune system antibodies attack and destroy insulin-producing cells in the pancreas. By young adulthood, people with type 1 diabetes require insulin injections to survive.

- **Guillain-Barre syndrome (GBS):** The immune system attacks the nerves controlling muscles in the legs and sometimes the arms and upper body. Weakness results, which can sometimes be severe. Filtering the blood with a procedure called plasmapheresis is the main treatment for GBS.

- **Chronic inflammatory demyelinating polyneuropathy (CIDP):** Similarly to GBS, the immune system also attacks the nerves in CIDP, but symptoms last much longer. About 30% of patients can become confined to a wheelchair if not diagnosed and treated early. Treatment for CIDP and GBS are essentially the same.

- **Psoriasis:** In psoriasis, overactive immune system blood cells called T-cells collect in the skin. The immune system activity stimulates skin cells to reproduce rapidly, producing silvery, scaly plaques on the skin.

- **Graves' disease (GD):** The immune system produces antibodies that stimulate the thyroid gland to release excess amounts of thyroid hormone into the blood (hyperthyroidism). Symptoms of GD can include bulging eyes as well as weight loss, nervousness, irritability, rapid heart rate, weakness, and brittle hair. Destruction or removal of the thyroid gland, using medicines or surgery, is usually required to treat GD.

- **Hashimoto's thyroiditis (HT):** Antibodies produced by the immune system attack the thyroid gland, slowly destroying the cells that produce the thyroid hormone. Low levels of thyroid hormone develop (hypothyroidism), usually over months to years. Symptoms include fatigue, constipation, weight gain, depression, dry skin, and sensitivity to cold. Taking a daily oral synthetic thyroid hormone pill restores normal body functions.

- **Myasthenia gravis (MG):** Antibodies bind to nerves and make them unable to stimulate muscles properly. Weakness that gets worse with activity is the main symptom of MG. Mestinon (pyridostigmine) is the main medicine used to treat MG.

- **Vasculitis:** The immune system attacks and damages blood vessels in this group of autoimmune diseases. Vasculitis can affect any organ, so symptoms vary widely and can occur almost anywhere in the body. Treatment includes reducing immune system activity, usually with prednisone or another corticosteroid.

Sidebar 3.4
Brain fog

What is Brain Fog?

Brain fog is a lay term to describe fluctuating mild memory loss that is inappropriate for a person's age. It may include forgetfulness, spaciness, confusion, decreased ability to pay attention, inability to focus, and difficulty in processing information. Remember that gradual cognitive decline from early adulthood is a fact of life.

What is the cause of brain fog?

Brain fog can occur in Sjögren's syndrome (SS). Recent scientific data show that longevity is associated with the successful management of chronic diseases, such as SS, not the absence of any disease! A major cause of cognitive dysfunction can be side effects of drugs and drug interactions, especially in patients over 65-70 years of age; but other factors that might be causing these symptoms should also be considered.

What can can be done about brain fog?

Managing one's lifestyle to optimize one's health and sense of well-being and developing a close working relationship with one's doctor(s) are recommended actions:

- Always reporting changes in cognition/memory and mood (depression, anxiety);
- Making sure the treating physician(s) knows about all the prescription and over-the-counter (OTC) medications being taken; and
- Inquiring about one's hormonal status, thyroid function, and blood pressure.

Additional actions include:

- Rejuvenating with sufficient sleep;
- Minimizing stress and anxiety:
- Setting realistic expectations;
- Planning ahead;
- Taking breaks throughout the day;
- Learning relaxation exercises and practicing them at regular intervals;
- Balancing work and leisure;
- Letting oneself laugh;
- Talking about feelings;
- Limiting multi-tasking and focusing on one task at a time;
- Reducing one's intake of caffeine and alcohol;
- Managing effectively musculoskeletal and joints pain;
- Exercising regularly;
- Adequately exercising physically, enhancing one's cognition/memory, and visiting the "brain spa";
- Training one's brain "If you don't use it, you will lose it";
- Boosting one's brain power;
- Continuing to work into retirement (part time); and
- Learning new skills; volunteering; and engaging in social and mentally-stimulating activities, and establishing new friendships and relationships.

4
Signs and symptoms of LD

LD can affect multiple body systems and produce a broad range of symptoms. Not everyone with LD has all of the symptoms, and many of the symptoms are not specific to LD, but can occur with other diseases, as well. (See the pictorial of Figure 4.1)

The incubation period from infection to the onset of symptoms is usually one to two weeks, but can be much shorter (days), or much longer (months to years). Lyme symptoms most often occur from May to September, because the nymphal stage of the tick is responsible for most cases (see Figure 2.5). Asymptomatic infection exists, but is thought quite rare, occurring in less than estimated diagnosed cases in the U.S. It may be much more common among those infected in Europe although there are no official estimates of such cases.

The symptoms of LD are largely due to the immune system's response to the presence of spirochetes in a particular organ or tissue with two possible additions: (a) the disruption of the hypothalamic-pituitary adrenal axis (the so-called "HPA-axis"), and (b) alterations in cell proliferation and death in brain tissue. In most affected areas in the U.S., symptoms are more likely to arise in the period between May and September due to the time delay between tick-bite and infection incubation and the spreading of the infection from the bite of a nymphal tick rather than an adult tick. However, the tick's tendency to bite humans does vary between different regions, and the type of tick hosts, such as deer, cattle, or sheep. On the other hand, ticks prevalent in Europe may be more likely to bite humans and spread infection in the adult stage rather than the nymphal stage. Where adult ticks spread the infection, the onset of symptoms will usually be felt in late Summer, Autumn, or even Winter as opposed to Spring and early Summer where nymphs are the main cause of tick bites and LD.

LD symptoms are associated with three developmental stages of the condition and it will be helpful to distinguish separately each stage in the onset of infection and the corresponding symptoms. The severity of LD symptoms tends to increase over time, as does the difficulty of LD treatment, although all stages of infection appear to respond well to short-term antibiotic treatment.

Early-localized infection

Early-localized (acute) infection can occur when the infection has not yet spread throughout the body. Only the site where the infection has first come into contact with the skin is affected. The initial sign of about 80% of Lyme infections is an *erythema migrans* (EM) rash at the site of a tick bite, often but not always near skin folds, such as the armpit, groin, the back of the knee, on the trunk, under clothing straps, or in children's hair, ear, or neck. Most people who get infected do not remember seeing a tick or the bite.

Figure 4.1 – Pictorial illustrating the signs and symptoms of Lyme disease

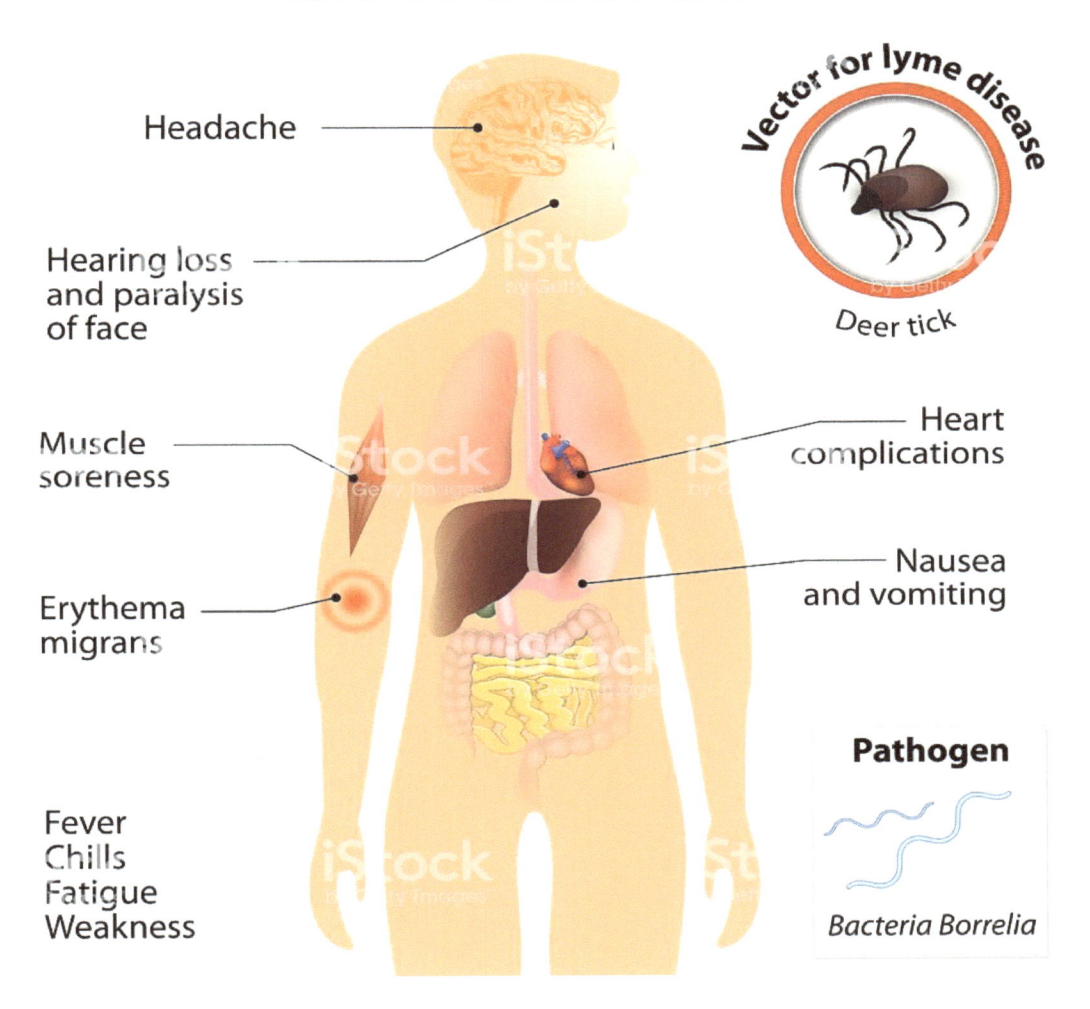

The rash appears typically one or two weeks (range 3-12 days) after the bite and expands 2-3 cm per day up to a diameter of 5-70 cm (median 16 cm). It is usually circular or oval, red or bluish, and may have an elevated or darker center. In 19% of cases in endemic areas of the U.S. and about 79% of cases in Europe, the rash gradually clears from the center toward the edges, possibly forming a "bull's eye" pattern. It may feel warm but usually is not itchy, rarely tender or painful, and takes up to four weeks to resolve if untreated. An expanding rash is an initial sign of about 80% of Lyme infection as illustrated by the photograph in Figure 4.2 of the upper-right arm of an infected woman. While this rash is observed in about 80% of LD patients, it is also observed in other illnesses not caused by *Borrelia*

burgdorferi infection, such as southern tick–associated rash illness (STARI), which has not yet been tied to a specific pathogen. It will generally wane over a two- to four-week period. 70% of all patients who do present with LD never recall such a rash.

Figure 4.2 – "Classic" erythema *migrans* rash characteristic of a Lyme infection

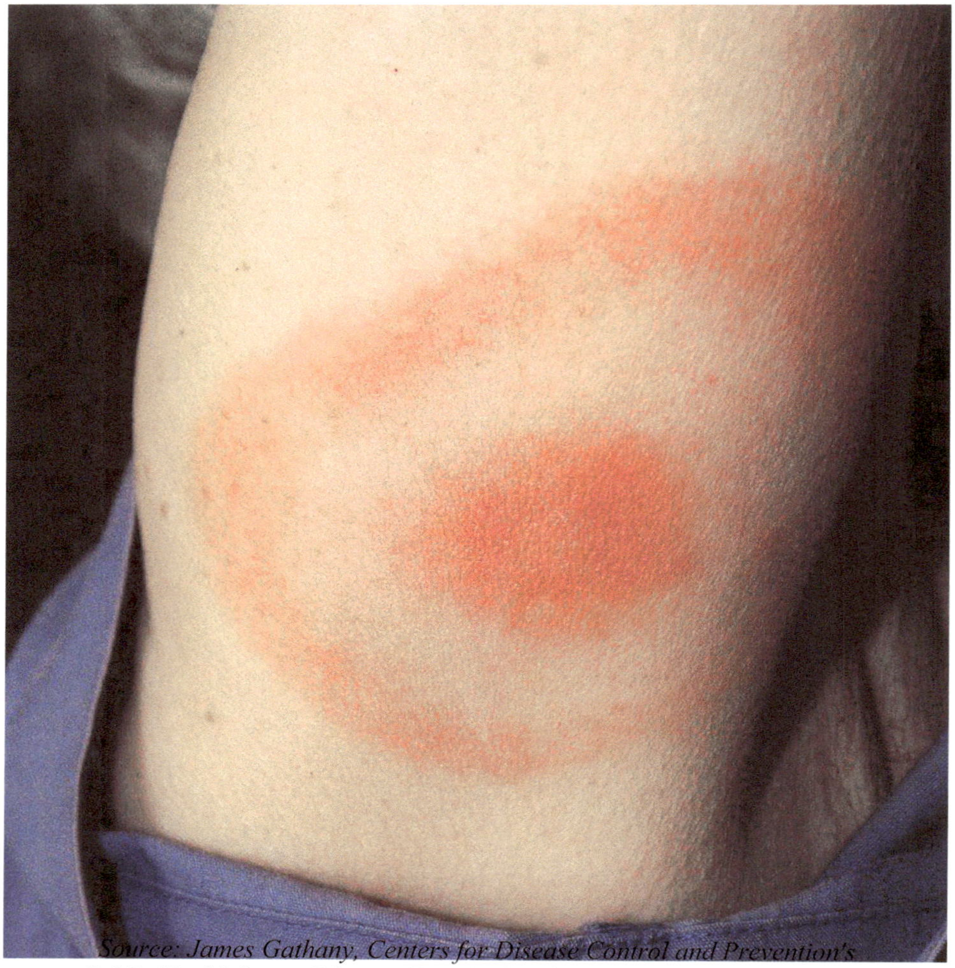

Source: James Gathany, Centers for Disease Control and Prevention's Public Health Image Library (PHIL), identification number #9875 (2007).

The EM rash is often accompanied by symptoms of a viral-like illness, including fatigue, headache, body aches, fever, and chills, but usually not nausea or upper-respiratory problems. These symptoms may also appear without a rash, or linger after the rash disappears. Lyme can progress to later stages without these symptoms or a rash. If left untreated, infection can spread to the joints, the heart, and the nervous system.

LD patients who are diagnosed early, and receive proper antibiotic treatment, usually recover rapidly and completely. A key component of early diagnosis is recognition of the characteristic LD *erythema migrans* rash.

People with high fever for more than two days or whose other symptoms of viral-like illness do not improve despite antibiotic treatment for LD, or who have abnormally low-levels of white or red cells or platelets in their blood, should be investigated for possible co-infection(s) with other tick-borne diseases such as those mentioned under the earlier acronyms (AB3CE) and (AC2EMPQ).

The above symptoms are summarized in Table 4.1.

Table 4.1 – Signs and symptoms of early-localized infection (3-30 days after tick bite)

Characteristics	Description
Feature	Local. Not yet spread to entire body
Signs *Erythema migrans* (EM) rash observed in ~ 70-80% of infected persons and in other infections not caused by *Borrelia burgdorferi*	o Occurs at the site of a tick bite after delay of 3-10 days (average ~ 7 days) o Expands 2-3 cm/day up to 5-70 cm/day (median 16 cm/day) o May feel warm at the touch, but is rarely itchy or painful o Sometimes clears as it enlarges, resulting in a target or "bull's eye" appearance (19% of cases in U.S., 79% of cases in Europe) o May appear in any area of the body o Gradually clears and resolves over 4 weeks, if left untreated
Symptoms May appear without a rash or linger after the rash has subsided	Viral-like illness: o Aches: body, head, joints, muscles o Chills o Fatigue o Fever o Swollen lymph nodes (may occur in the absence of EM rash) Note: Lyme can progress to later stages even in the absence of above symptoms
If diagnosed and treated early	Quick and complete recovery after antibiotic treatment
If untreated	Infection can spread to: o Heart o Joints o Nervous system
Caution Need for investigation of co-infections	o Abnormally low levels of white or red blood cells or platelets in the blood o Persistent fever for more than 2 days, or o Viral-like symptoms do not improve despite antibiotics

Early-disseminated infection

Within days to weeks after the onset of local infection, the *Borrelia* bacteria may spread through the lymphatic system or bloodstream. In 10-20% of untreated cases, EM rashes develop at sites across the body that bear no relation to the original tick bite. Transient muscle and joint pains are also common. Table 4.2 summarizes the signs and symptoms of early-disseminated infection (days to months after the tick bite).

In about 10-15% of untreated people, Lyme causes neurological problems known as *neuroborreliosis* (NB). Early NB typically appears 4–6 weeks (range 1–12 weeks) after the tick bite and involves some combination of cranial neuritis, lymphocytic meningitis, radiculopathy and/or mononeuritis multiplex:

- **Cranial neuritis (CN):** This is an inflammation of cranial nerves. When due to Lyme, it most typically causes facial palsy (FP), impairing blinking, smiling, and chewing in one or both sides of the face. It may also cause intermittent double vision.

- **Lymphocytic meningitis (LM):** It causes characteristic changes in the cerebrospinal fluid (CSF). For several weeks, it may be accompanied by variable headache and usually, but less commonly, mild meningitis signs such as inability to flex the neck fully and intolerance to bright lights, but typically no or only very low fever. In children, partial loss of vision may also occur.

- **Radiculopathy:** It is an inflammation of spinal nerve roots that often causes pain and less often weakness, numbness, or altered sensation in the areas of the body served by nerves connected to the affected roots, for example, the limb(s) or part(s) of the trunk. The pain is often described as unlike any other previously felt, excruciating, migrating, worse at night, rarely symmetrical, and often accompanied by extreme sleep disturbance.

- **Mononeuritis multiplex (MNM):** This is an inflammation causing similar symptoms in one or more unrelated peripheral nerves. Rarely, early *neuroborreliosis* may involve inflammation of the brain or spinal cord with symptoms such as confusion, abnormal gait, ocular movements, slurred speech, impaired movement, impaired motor planning, or shaking (a Parkinsonian syndrome).

Statistics for untreated cases

The following statistics have been compiled for untreated Lyme disease cases:

- *In North America:* Facial palsy (FP) is the typical early *neuroborreliosis* presentation, occurring in 5%-10% of untreated people. In about 75% of cases, it is accompanied by lymphocytic meningitis (LM). Lyme radiculopathy is reported half as frequently, but many cases may be unrecognized.

- *In Europe:* In adults, the most common presentation is a combination of LM and radiculopathy known as Bannwarth syndrome (BS), accompanied in 36%-89% of cases by facial palsy (FP). In this syndrome, radicular pain tends to start in the same body region as the initial EM rash, if there was one, and precedes possible FP, and other impaired movement. In extreme cases, permanent impairment of motor or sensory function of the lower limbs may occur. In European children, the most common manifestations are FP (in 55% of cases), other CN, and LM (in 27% of instances). Another skin condition, found in Europe but not in North America, is borrelial lymphocytoma (BL), a purplish lump that develops on the ear lobe, nipple, or scrotum.

- ***In the U.S. and Europe:*** In about 10% of untreated cases in the U.S. and 0.3%-4% of untreated cases in Europe, typically between June and December, about one month (range 4 days-7 months) after the tick bite, the infection may cause heart complications known as Lyme carditis (LC). Symptoms may include heart palpitations (in 69% of people), dizziness, fainting, shortness of breath, and chest pain. Other symptoms of LD may also be present, such as EM rash, joint aches, facial palsy, headaches, or radicular pain. In some people, however, carditis may be the first manifestation of LD. Lyme carditis in 19%-87% of people adversely impacts the heart's electrical conduction system, causing atrioventricular block that often manifests as heart rhythms that alternate within minutes between abnormally slow and abnormally fast. In 10%-15% of people, Lyme causes myocardial complications such as cardiomegaly, left ventricular dysfunction, or congestive heart failure (CHF).

Table 4.2 – Signs and symptoms of early disseminated infections
(days in months after tick bite)

Characteristics	Description
Feature Spread through the lymphatic system or the blood stream	10-15% of untreated cases: o Additional EM rashes develop to sites outside the tick bite region o Episodes of dizziness or shortness of breath o Joints pain o Neck stiffness o Severe headaches o Shooting pains, numbness or tingling in the hands, or feet o Transient muscle pains
Signs Neurological problems - *Neuroborreliosis*	10-15% of untreated cases: o *Neuroborreliosis* appears within 4-6 weeks (range 1-12 weeks) after the bite o Some combination of cranial neuritis, lymphocytic meningitis, radiculopathy, and/or mononeuritis multiplex
Symptoms	o Arthritis with severe joint pain and swelling, particularly in the knees and other large joints o Intermittent pain in tendons, muscles, joints, and bones ***Central nervous system:*** o Inflammation of the brain and spinal cord o Nerve pain ***Cranial neuritis:*** o Facial palsy (loss of muscle tone or droop on one or both sides of the face) o Impaired blinking, smiling, and chewing in one or both sides of the face o Intermittent double vision

Lyme disease

	Lymphocytic meningitis: o Cerebrospinal fluid changes o Mild meningitis (neck flexing inability; intolerance to bright light; no or very low fever; partial vision loss in children) o Variable headaches ***Mononeuritis multiplex:*** o Inflammation in one or more peripheral nerves, brain, spinal cord o Parkinsonian movement disorder ***Radiculopathy:*** o Numbness and/or o Pain o Spinal nerve roots inflammation o Weakness
If untreated	o Carditis (4%-10% of cases) in about 1 month (range: 4 days- 7 months) o Facial palsy (5%-10% of cases) o Lymphocytic meningitis (75% of cases) o Radiculopathy (50% of cases; some unrecognized)
Caution	***Lyme carditis:*** o Chest pain o Dizziness o EM rash o Facial palsy o Fainting o Headaches o Heart palpitations or irregular heart beats (69% of cases) o Joints aches o Radicular pain o Shortness of breath ***Myocardial complications:*** o Cardiomegaly o Congestive heart failure (CHF) o Left ventricular dysfunction

Figure 4.3 illustrates facial palsy (loss of muscle tone or droop on one or both sides of the face) and Figure 4.4 illustrates a swollen knee.

Late-disseminated infection

This stage is also referred to as "late persistent infection". After several months, untreated or inadequately treated people may go on to develop chronic symptoms that affect many parts of the body, including the joints, the nerves, the brain, the eyes, and the heart.

Lyme arthritis (LA) occurs in up to 60% of untreated people, typically starting about six months after infection. It usually affects only one or a few joints, often a knee or possibly the hip, other large joints, or the temporomandibular joint (TMJ). There is usually large joint effusion and swelling, but only mild or moderate pain. Without treatment, swelling and pain typically resolve over time but periodically return. Baker cysts may form and rupture. In some cases, joint erosion occurs. Chronic neurologic symptoms occur in up to 5% of untreated people. A peripheral neuropathy or polyneuropathy may develop, causing abnormal sensations such as numbness, tingling or burning starting at the feet or hands and over time possibly moving up the limbs. A test may show reduced sensation of vibrations in the feet. An affected person may feel as if wearing a stocking or glove without actually doing so.

Figure 4.3 – Illustrating facial palsy

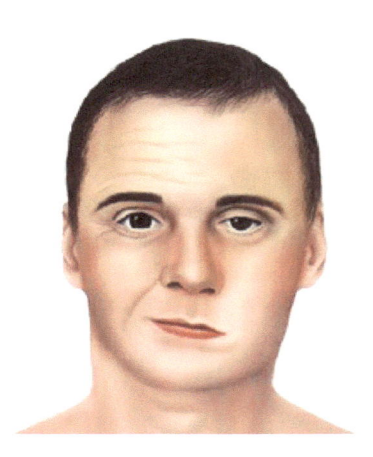

Figure 4.4 – Illustrating a swollen knee

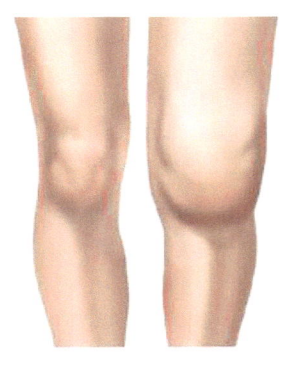

A neurologic syndrome called Lyme encephalopathy (LE) is associated with subtle memory and cognitive difficulties, insomnia, a general sense of feeling unwell, and changes in personality. However, problems such as depression and fibromyalgia are as common in people with LD as in the general population.

Lyme can cause a chronic encephalomyelitis that resembles multiple sclerosis (MS). It may be progressive and can involve cognitive impairment, brain fog, migraines, balance issues, weakness in

Lyme disease

the legs, awkward gait, facial palsy, bladder problems, vertigo, and back pain.

In rare cases, untreated LD may cause frank psychosis, which has been misdiagnosed as schizophrenia or bipolar disorder. Panic attacks and anxiety can occur. Delusional behavior may also be seen, including somatoform delusions sometimes accompanied by a depersonalization or derealization syndrome, where the people begin to feel detached from themselves or from reality. *Acrodermatitis chronica atrophicans* (ACA) is a chronic skin disorder observed primarily in Europe among the elderly. ACA begins as a reddish-blue patch of discolored skin, often on the backs of the hands or feet. The lesion slowly atrophies over several weeks or months, with the skin becoming first thin and wrinkled and then, if untreated, completely dry and hairless.

Table 4.3 – Signs and symptoms of late-disseminated infection

Characteristics	Description
If untreated or inadequately treated Wide range of symptoms depending on the stage of the disease	*Chronic symptoms* that affect many parts of the body: o Arthritis o Brain o Eyes o Facial paralysis o Fever o Heart o Joints o Nerves o Rash *Chronic encephalomyelitis:* o Awkward gait o Back pain o Balance issues o Bladder problems o Brain fog o Cognitive impairment o Facial palsy o Migraines o Resembles multiple sclerosis (may be progressive) o Vertigo o Weakness in the legs *Chronic neurologic symptoms* (up to 5% of cases) o Peripheral neuropathy or polyneuropathy *o Lyme arthritis* (60% of untreated cases; 6 months after infection) o Large joints effusion and swelling (resolves over time but periodically returns) o Baker cysts o Joints erosion *Lyme encephalopathy:* o Changes in personality o Depression o Fibromyalgia o General sense of feeling unwell

Lyme disease

	o Insomnia o Subtle memory and cognitive difficulties
Rare untreated cases	***Psychosis:*** o Anxiety o Delusional behavior (somatoform delusions, derealization, depersonalization) o Panic attacks ***Acrodermatitis chronica atrophicans:*** (mostly in Europe) o Chronic skin disorder

Untreated late-disseminated LD can produce a wide range of symptoms, depending on the stage of infection. Such symptoms are summarized in Table 4.3 above.

Take-home points

- Lyme disease can affect multiple body systems and produce a broad range of symptoms. Not every affected person has all of the symptoms. Many of the symptoms are not specific to the disease but can occur with other diseases as well.

- The incubation period from infection to the onset of symptoms is usually one to two weeks, but can be much shorter or much longer. Symptoms most often occur from May to September, because the nymphal stage of the tick is responsible for most cases.

- Asymptomatic infection exists, but occurs in less than 7% of infected individuals in the U.S. It may be much more common among those infected in Europe.

- There are three distinct phases for the disease: early-localized infection, early-disseminated infection, and late-disseminated infection that have been conveniently summarized in Tables 4.1-4.3.

- Early-localized infection can occur when the infection has not yet spread throughout the body. It is characterized by a rash (called *erythema migrans)* at the site of a tick bite, but often near skin folds. The rash, a key component of early diagnosis, appears typically 1-2 weeks after the bite and expands 2-3 cm per day. It gradually clears from the center toward the edges, possibly forming a "bull's eye" pattern. However, while this rash is observed in about 80% of cases, it is also observed in other illnesses. The rash is often accompanied by symptoms of a viral-like illness (fatigue, headache, body aches, fever, and chills) but these symptoms may also appear without a rash. If left untreated, infection can spread to the joints, the heart, and the nervous system.

- Lyme patients who are diagnosed early, and receive proper antibiotic treatment, usually recover rapidly and completely.

- People with high fever for more than two days or whose other symptoms of viral-like illness do not improve despite antibiotic treatment, or who have abnormally low-levels of white or red cells or platelets in their blood, should be investigated for possible co-infections with other tick-borne diseases.

- Facial palsy is the typical early-disseminated infection. In ~ 75% of cases, it is accompanied by lymphocytic meningitis and radiculopathy, possibly carditis in about 4-10% of untreated cases (heart palpitations, dizziness, fainting, shortness of breath, and chest pain, alternating heart rhythms, cardiomegaly, left ventricular dysfunction, or congestive heart failure). Other symptoms may also be present (joint aches, facial palsy, headaches, or radicular pain).

- After several months, untreated or inadequately treated people may go on to develop late-disseminated infection as chronic symptoms that affect many parts of the body (joints, nerves, brain, eyes, and heart). Arthritis occurs in up to 60% of untreated people (few joints, knee, hip, large joints, temporomandibular joint). Without treatment, swelling and pain typically resolve over time but periodically return. Baker cysts may form and rupture.

- Chronic neurologic symptoms occur in up to 5% of untreated people (peripheral neuropathy or polyneuropathy, encephalopathy, depression, and fibromyalgia). They are as common in people with LD as in the general population. Chronic progressive encephalomyelitis resembling multiple sclerosis may also develop (cognitive impairment, brain fog, migraines, balance issues, weakness in the legs, awkward gait, facial palsy, bladder problems, vertigo, and back pain).

- Psychosis may develop in rare cases (panic attacks, anxiety, delusional behavior including somatoform delusions).

- *Acrodermatitis chronica atrophicans,* a chronic skin disorder, may be observed primarily in Europe among the elderly.

- For the interested readers, some selected recent epidemiological and surveillance data where Lyme Disease occurs are presented and illustrated in Chapter 9 by State, age group, and gender.

5

Getting the correct LD diagnosis

Testing and monitoring is a complicated multi-step process that is still evolving as we learn more about the disease and new diagnostic tools are devised. It is important for patients to monitor their progress, otherwise, how will anyone know if they are improving or at risk of relapse? It is equally important to have the proper tests, understand how they work, and determine the best possible pathway for modifying treatment to attain long-term health. In this process, establishing a comprehensive list of infections associated with LD may, in fact, help patients receive a proper diagnosis in order to administer the much needed comprehensive treatments they deserve.

On the importance of differentiating the underlying bacterial strain

The spirochetal agent that causes LD is the *Borrelia* genus of bacteria, with *Borrelia burgdorferi sensu lato* (Bbsl) being the broad categorization of bacteria related to the condition. Specific strains of *Borrelia* bacteria within this category are prevalent in different geographical locations. *Borrelia burgdorferi sensu stricto* (Bbss) is held responsible for all cases of Lyme disease in North America and some cases in Europe. The *Borrelia afzelii* and *Borrelia bavariensis* spirochetes are found in Europe along with *Borrelia garinii*, which is also prevalent in Asia. There are several other strains which have an association with LD although infectious cases are rare and unconfirmed for most of these. Hybridization has been found in *Borrelia* bacteria and a large volume of research has already accumulated documenting the presence and prevalence of new variations of the genus in varying locations around the world.

The importance of the particular bacterial strain that is responsible for the infection is only just being revealed with researchers observing patterns of LD symptoms associated with each type of *Borrelia* bacteria. In Europe, where the *Borrelia afzelii* bacteria is the common cause of infection, arthritis as a symptom of LD is much less likely to develop than in the US. where Bbss often leads to this symptom. *Borrelia afzelii* is more often associated with the development of the chronic skin condition *acrodermatitis chronica atrophicans* (ACA) and *Borrelia garinii* with neurological symptoms. Thus, knowing the prevalent strains of spirochete in an area can allow the population to be on the lookout for specific symptoms in order to catch LD in its localized early (acute) stage rather than as it disseminates and becomes harder to treat. Early LD symptoms can be easily overlooked and those who are aware of the risk of LD in their communities are usually more likely to seek early medical attention after observing more subtle symptoms of infection.

Issues with current diagnostic methodology

The diagnosis of LD is based primarily on objective signs of a known exposure, clinical findings, and supportive serologic (blood) testing. The problem in current testing for LD is the high likelihood of receiving a false negative. A false negative occurs when a test produces results indicating that a disease is not present when, in reality, it is. Western blot and ELISA (Enzyme-Linked Immuno-Sorbent Assay) are the standard testing methods used to diagnose LD, but these older tests may have a high chance of producing false negatives because of the method they employ to produce their results.

Not only old and outdated, Western blot and ELISA can be highly flawed. Unfortunately, they represent the only option for most doctors to test for LD. The tests look for the presence of certain antibodies (AB) to produce a positive result when searching for LD. However, LD patients are often immune-compromised and their bodies may not be producing the antibodies necessary to conclusively identify the existence of LD. In addition, these tests do not quantify their results so patients and doctors cannot identify the number of copies each infection type present has produced and, therefore, are unable to provide conclusive data for therapeutic removal of the infection. LD patients are often caught in a vicious cycle of immune depression that began with the initial infection. It is important to restate that when bitten by a tick, together with LD, co-infections that the tick may carry or enable are also transferred with *Borrelia burgdorferi*, the main bacterial spirochete known for causing LD. Once *Borrelia* and these co-infections enter the patient's body, they release a multitude of endotoxins, neurotoxins, and biotoxins. Among many other negative actions, the toxins confuse the immune system and cause it to potentially attack its own cells, resulting in autoimmune-like symptoms and chronic inflammation throughout the brain and body. The confusion caused by toxins, and other depressors of the immune system, makes the patient's body more susceptible to opportunistic secondary fungal, viral, parasitic, and other bacterial infections, further impairing the patient's immune system. With the immune system severely impaired by Lyme and these other infections, the traditional tests may not be accurate for diagnosis. To make matters worse, the initial infection, co-infections, biotoxins, endotoxins, mycotoxins, neurotoxins, autoimmune attacks, and opportunistic secondary infections... can produce a multitude of symptoms making it difficult to recognize that Lyme may be a possibility. It is the combination of these multiple infections with other complications that contribute to the patients' debilitating set of physical and neurological symptoms described. This all combines to make a very confusing and frustrating experience for the patient and the treating doctor alike.

Correctly identifying the infectious load

Only in-depth, up-to-par laboratory testing can identify the above several infections. Correctly identifying and vigorously treating all of them is the key to producing lasting results. However, as previously indicated, blood tests are often negative in the early stages of the disease. Further, testing of individual ticks is not typically useful as it is their combined interacting effects (not their individual effects) that are relevant. The common tests (Western Blot and ELISA) used to confirm a Lyme diagnosis can report incorrect results 50% of the time (see below). Indeed, erroneous test results have been widely reported in both early and late stages of the disease. They can be caused by several factors such as antibody cross-reactions from other infections including Epstein-Barr virus (EBV) and cytomegalovirus (CMV) as well as herpes simplex virus (HSV).

To complicate matters, Lyme patients with neurological symptoms are often misdiagnosed with one or more of the other neurological diseases discussed or, even worse, their symptoms are dismissed altogether. The struggle to find the right diagnosis is extremely draining for patients who are already facing depression, brain fog, memory loss, tremors, and other crippling symptoms.

(Note: An established and well-known clinical facility claims to have developed an early version of a "proprietary PCR (polymerase chain reaction) diagnosis protocol", which is still in the research and developmental stage, allowing its physicians to not only accurately diagnose LD, but be able to monitor the several infection and co-infection levels throughout treatment. However, being proprietary, it is difficult to gauge that claim pending disclosure, peer-review, independent confirmation, and acceptance of that protocol. Further, PCR is itself outdated and inferior to genomic testing. See below more detailed comments on PCR.)

The four basic principles of the diagnosis

The four basic principles of the diagnosis are:

- **History of possible exposure to infected ticks;**

- **Signs and symptoms observed;**

- **Objective physical findings** (such as *erythema migrans* (EM) rash, facial palsy, arthritis, etc.); and possibly

- **Supportive laboratory tests.**

In the diagnosis process, it is useful to distinguish the three instances discussed in Chapter 4: (1) localized (or early) disease, (2) early-disseminated disease, and (3) late-disseminated disease.

Case of localized (or early) LD

People with symptoms of early LD should have:

- A total body skin examination for EM rashes and

- Be inquired if there was a rash in the past 1–2 months.

Presence of an EM rash and recent tick exposure (i.e., being outdoors in a likely tick habitat where Lyme is common, within 30 days of the appearance of the rash) are sufficient for Lyme diagnosis; no laboratory confirmation is needed or recommended. Unfortunately, most people who get infected do not remember a tick or a bite, and the EM rash need not look like a bull's eye (most EM rashes in the U.S. do not) or be accompanied by any other symptoms. In the U.S., Lyme is most common in the New

England and Mid-Atlantic states and parts of Wisconsin and Minnesota, but it is expanding into other areas. Several bordering areas of Canada also carry high Lyme risk.

Table 5.1 – Diagnostic methodology: Localized (or early) disease

Principle	Test	Advantages	Shortcomings
Recent history and total body skin exam for rash	Evidence of exposure, rash, and symptoms	o Sufficient for diagnosis o No laboratory tests required	**Only for early Lyme Disease**
	Lack of exposure or rash	Laboratory tests required for antibody (AB) presence	o Positive AB test does not prove active infection o Can confirm suspected infection because of symptoms and objective findings Note: 5%-20% of the normal population have AB against Lyme

In the absence of an EM rash or a history of tick exposure, Lyme diagnosis depends on laboratory confirmation. The bacteria that cause LD are difficult to observe directly in body tissues. They are also difficult and too time-consuming to grow in the laboratory. The most widely used tests look instead for the presence of antibodies against those bacteria in the blood. However, a positive antibody test result does not by itself prove active infection. It can, however, confirm an infection that is suspected because of symptoms, objective findings, and history of tick exposure in a person. Because as many as 5%-20% of the normal population have antibodies against Lyme, people without history and symptoms suggestive of LD should not be tested for Lyme antibodies because a positive result would likely be false, possibly causing unnecessary treatment. This is summarized in Table 5.1 above.

For those patients who do get diagnosed correctly, the standard-of-care involves a simple course of antibiotics.

People who suspect they may have LD should seek out specialty laboratories for a more accurate diagnosis. One such specialty laboratory is IGeneX. IGeneX looks at samples of the patient's blood under a microscope. The inspection allows them to visualize and identify so-called "bands" of the spirochetes involved with LD. IGeneX also runs a separate co-infection panel to help diagnose the other infections that can be present with LD. It is important to mention that this test is only a starting point. There is a number of other specialized laboratory tests that must be run to aid in a proper and complete diagnosis. These are necessary to create a full medical blue print for LD and all the co-infections that can be present. The multitude of symptoms associated with LD makes it hard to diagnose it based on symptoms alone. But, with proper diagnosis and treatments tailored to all the co-infections, patients can typically rid themselves of most of the symptoms that are associated with LD and receive more long-term benefits from treatment.

Case of early-disseminated LD

In some cases, when history and signs and symptoms are strongly suggestive of early-disseminated LD, empiric treatment may be started and re-evaluated as laboratory test results become available. The (U.S.) Centers for Disease Control & Prevention (CDC&P) has recommended the following two-tiered protocol whose reliability remains controversial. Tests for antibodies in the blood are ELISA and the Western blot, the former being the most widely used method for Lyme diagnosis.:

- A sensitive first test, either an enzyme-linked immunosorbent assay (ELISA) or an indirect fluorescent antibody test (IFAT), followed by

- The more specific Western immunoblot (WB) test to corroborate equivocal or positive results obtained with the first test.

High titers of either immunoglobulin G (IgG) or immunoglobulin M (IgM) antibodies to *Borrelia* antigens indicate disease, but lower titers can be misleading because the IgM antibodies may remain after the initial infection and IgG antibodies may remain for years (see Table 5.2 where I also mention additional tests). Patients who do not have access to DNA-sequencing may experience difficulties getting an accurate diagnosis for their LD under the standard form of testing.

Although the standard-of-care for correctly diagnosed LD patients is a simple course of antibiotics, that treatment may only be effective in some cases. Thus, there are many patients who will continue to experience ongoing resistance to antibiotics and continued symptoms (what was termed earlier Post-Treatment Lyme Disease Syndrome, PLDS). To repeat, PLDS does not define the full breadth of infections the patients may have, regardless of whether they have gone through standard-of-care treatment for Lyme *borrelia* or not. In fact, the Lyme *borrelia* may not always be present in a CLD patient as the term refers to a wide range of tick-borne plus other secondary infections and complications that the patient may be dealing with. This infective load and the complications associated with it vary greatly from patient to patient so that a more personalized treatment approach would often lead to an improved patient outcome. In particular, those patients that present with neurological Lyme need specialized care that is critical for their improvement. (Note that the CDC&P does not recommend either of the following tests: urine antigen tests; PCR tests on urine; immunofluorescent staining for cell-wall-deficient forms of *Borrelia burgdorferi*; and lymphocyte transformation tests.)

Table 5.2 – Diagnostic methodology: Early-disseminated disease
(History, signs, and symptoms are strongly suggestive)

Principle	Test	Advantages	Shortcomings
Titers of IgG or IgM Schedule: 2-4 weeks: IgM 4-6 weeks: IgG 6-8 weeks: IgM > 8 weeks: only IgM	**ELISA** (test for antibody (AB) and color changes)	o High titers indicate disease but lower titers can be misleading because IgG AB may remain for years o IgM AB may remain after the initial infection	o Initial sensitivity ~ 70% o Immune system not for insufficiently producing AB to allow for proper diagnosis o Better for locating infection, not monitoring

Lyme disease

		o Positive IgM and negative IgG indicate infection	overall health o Often produce false-negative results in immuno-compromised patients and patients on steroids or on suppressive medication for autoimmune disease o Valid only after 30 days o Less useful after AB treatment o Antiquated and outdated for definitively diagnosing and quantifying Lyme *borreliosis* o Fails to detect and accurately quantify the presence of any present co-infection(s)
	Western blot (if ELISA not specific enough)	o Detects specific amino-acid sequences in proteins o Identifies the protein and determines the correct treatment or AB	o Initial sensitivity ~ 94%-96% o Antiquated and outdated for definitively diagnosing and quantifying Lyme *borreliosis* o Fails to detect and accurately quantify any present co-infection(s)
Microscopy	**Dark-field microscopy**	o Probably one of the best tests o Visualizes the spirochete when it emerges from hiding	o Spirochete is only seen in ~ 40% of cases o Out of the host (*borrelia*) changes shape and hides intracellularly immediately o Method is not 100% accurate
Cytokine IFN-g	Ispot	Secreted by patient's T-cells	o Better specificity than the Western blot
OspA antigens	**OspA** (uses nanotrap particles for detection)	o Antigens shed live bacteria in urine o Promising	o In development
Polymerase chain reaction (PCR)	PCR	o Detects the genetic material (DNA) of the LD spirochete o Much faster than laboratory culture o May be considered when intrathecal AB-producing test results are suspected of being falsely negative	o Susceptible to false positive results o Often shows false negative results o Recommended only in special cases (for example, Lyme arthritis)
Genomics (a "gold standard" on the horizon)	**DNA, RNA sequencing** (from patient's blood, urine, mucus, stools)	o More useful than current tests (including PCR) o Provides conclusive and quantifiable data on total infectious load	o More data and validation are still needed

Lyme disease

		o Leads to better diagnosis and treatment while providing a more accurate way to track the progress of treatment o Accurately tracks treatment progression o More appropriate personalized treatment o Allows selection of best treatment drugs discarding non-drug targets	
Neurologic	*Neuroborreliosis* (Lumbar puncture and CSF analysis of pleocytosis and intrathecal AB production)	o In Europe o In the U.S.: Confirms a diagnosis of *neuroborreliosis* if positive. Does not exclude *neuroborreliosis* if negative	o American guidelines consider CSF analysis optional when symptoms are confined to the PNS (for example, facial palsy without overt meningitis symptoms)
Cardiologic	**Carditis**	Uses EKG	Not done because of associated risk

Key: *AB: Antibodies; CNS: Central Nervous System; CSF: Cerebrospinal Fluid; EKG: Electrocardiogram; ELISA: Enzyme-Linked Immuno-Sorbent Assay; Ig: Immuno-globulin; LD: Lyme Disease; PCR: Polymerase Chain Reaction; PNS: Peripheral Nervous System*

Some brief remarks on the tests in Table 5,2 can be found below:

- **ELISA:** This standardized type of laboratory blood test is the most common. It is considered to be the correct way to diagnose LD. ELISA is a type of wet-lab test that uses antibodies and color changes to identify a substance, in this case, LD. What we are looking for in this instance are antigens (a cell's identifying feature, like a microscopic caller ID) from the *Borrelia* bacteria. When laboratories find these antigens, they attach a specific antibody (a blood protein used to identify bacteria and viruses) which combines with the *Borrelia* antigen. Next, the enzyme's substrate (the surface on which a cell feeds itself) is added, producing a detectable signal, usually a change in color. This helps the laboratory define which foreign bodies are in the bloodstream. These standardized tests rely on the body's immune status and its production of antibodies to detect the disease. However, because LD is so evasive, often times the immune system is not producing or not producing sufficient numbers of these antibodies to allow for proper diagnosis. Also, as already said earlier, they can often produce false-negative results.

  ELISA is a sensitive test (initial sensitivity about 70%) that is performed first. It it is positive or equivocal, then, the more specific Western blot (94%-96% specificity for people with clinical symptoms of early LD) is run. When an EM rash first appears, because the immune system takes some time to produce antibodies in quantity, antibodies usually cannot yet be detected; therefore, antibody confirmation at that time has no diagnostic value and is not recommended. Up to 30 days after suspected Lyme infection onset, infection can be confirmed by detection of IgM or IgG antibodies. Even though the CDC&P recommends ELISA as the first line of testing, it can provide a false negative in patients with weakened immunity or on medications such as steroids. It is not uncommon for LD patients to be originally diagnosed with an autoimmune disease or fibromyalgia and, therefore, be given suppressive medication to manage symptoms instead of truly treating the disease.

The following testing schedule is usually applied:

At 2–4 weeks: IgM antibodies can first be detected but they usually collapse 4-6 months after infection. Immunoglobulin M is the first antibody to respond to initial antigen exposure. It is larger than IgG and the biggest antibody in the human circulatory system. When EM is detected in a LD patient, IgM identifies it but only during the first four weeks of a bite, if the rash is even present. It is important to note that only 25%-30% of Lyme patients remember or experience a rash. If IgM is not used to detect Lyme, it can lead to a misdiagnosis of LD. Additional tests may be more helpful, such as IgG. However, usually, by the time patients come to recognize they have the LD complex, they are beyond IgM testing.

At 4–6 weeks: IgG antibodies can next be detected and can remain detectable for years. Immunoglobulin G is an antibody isotype, making up 75% of immunoglobulins (proteins that work as antibodies) that are present in the bloodstream. Because IgG is so plentiful, it is the main factor in controlling infection. Its levels indicate a patient's immune status to particular pathogens (microorganisms that can cause disease). However, once positive, this is not a good tool to monitor progression or improvement. The test can stay positive, which does not provide clear clinical information in the monitoring of the patient. This is really more of an initial diagnostic tool for patients that show signs and symptoms of CLD.

At 6–8 weeks: Both IgM and IgG peak. The overall sensitivity is only 64%, although this rises to 100% in the subset of people with disseminated symptoms, such as arthritis.

After 8 weeks: It is recommended that only IgM antibodies be considered.

Note that the combination of a positive IgM and a negative IgG test result suggests an early infection, especially if confirmed several weeks later by a positive IgG test result.

After antibiotic treatment, antibody tests become less useful. People treated with antibiotics when they have an EM rash often subsequently test negative for Lyme antibodies, whether treatment was successful or instead Lyme goes on to cause further complications. People treated later usually test positive before and after treatment, regardless of treatment success or failure. This suggests that better diagnostic tests are needed.

The overall rate of false positives is low, only about 1%-3%, in comparison to a false-negative rate of up to 36% in the early stages of infection using the two-tiered testing.

- **Western blot ;** The problem with using IgG, IgM and ELISA is that they are better for locating infection, not monitoring overall health. The Western blot is a widely accepted analytical technique that detects specific amino-acid sequences in proteins. There are hundreds of thousands of different proteins, but once the protein is identified, one can determine the correct treatment or antibody.

- **Dark-field microscopy test:** The dark-field microscopy test with silver nitrate stain is probably the best test, though the spirochete is only seen in about 40% of cases. Unfortunately, it is beset by two problems. First, when taken out of the host during a blood test, the *borrelia* changes shape and immediately hides intracellularly. Second, the method is not always 100% accurate. However, improvements of this method are being worked on by creating a medium (serum) in which *Borrelia* can live up to eight weeks so that, upon retesting the blood, the spirochete reemerges from hiding.

- **Ispot Lyme Test:** This test has a better specificity than the Western blot test when testing for *Borrelia*. It measures the cytokine IFN-g secreted by the patient's T cells.

Other forms of laboratory testing: Other forms of laboratory testing for LD are available, some of which have not been adequately validated. Outer specific protein A (OspA) antigens shed by live *Borrelia* bacteria into urine is a promising technique being studied. For their detection, the use of nanotrap particles is being looked at. These other tests include:

- **Polymerase chain reaction (PCR) tests:** These tests have also been developed to detect the genetic material (DNA) of the LD spirochete. Whereas serologic studies only test for antibodies of *Borrelia,* culture or PCR is the current means for detecting the presence of the organism. PCR has the advantage of being much faster than laboratory culture. However, PCR tests are susceptible to false positive results, e.g. by detection of debris of dead *Borrelia* cells or specimen contamination. Even when properly performed, PCR often shows false negative results because few *Borrelia* cells can be found in blood and CSF during infection. Hence, PCR tests are recommended only in special cases, e.g. diagnosis of Lyme arthritis, because it is a highly sensitive way of detecting *OspA* DNA in synovial fluid. Although sensitivity of PCR in CSF is low, its use may be considered when intrathecal antibody production test results are suspected of being falsely negative, e.g. in very early (< 6 weeks) *neuroborreliosis* or in immunosuppressed people.

  While PCR can detect bacterial DNA in some patients, unfortunately, this is also not helpful as a test of whether the antibiotics have killed all the bacteria. Studies have shown that DNA fragments from dead bacteria can be detected for many months after treatment. The studies have also shown that the remaining DNA fragments are not infectious. Positive PCR test results are analogous to a crime scene: just because a robbery occurred and the robber left his/her DNA, it does not mean that the robber is still in the house. Similarly, just because DNA fragments from an infection remain, it does not mean the bacteria are alive or viable.

- **Genomic testing: A more accurate test on the horizon?** Nothing can describe the frustration and powerlessness of being sick and in pain with no clear reason why. LD is a debilitating and painful disease that all too often goes misdiagnosed or produces false negatives on the traditional ELISA and Western blot tests. However, there is hope on the horizon through the use of modern genomic technology. This approach is more useful than current tests, including the

PCR technique. Through this novel approach, the presence of LD and other specified co-infections can be more conclusively identified and quantified. This may greatly help in prescribing the appropriate personalized treatment and accurately track its progress for each patient. This new test may also aid in selecting the best drugs to target each organism in a patient's CLD.

Genomic testing is based on genomic information from the patient's blood, urine, mucus, and stool to derive important DNA and RNA information without having to synthetically amplify the genes for detection as required by the PCR method. Not only allowing for a better diagnosis, it also paves the way for a personalized, comprehensive treatment plan and quantifies the amount of infections present. At this time, more data and validation are still needed to bring this test to patients, but we hope it may become the "gold standard" for testing and treating tick-borne infections as well as a powerful tool in helping patients with CLD. Cross-referencing the patient's test with known data on the DNA and RNA sequences of diseases will reveal the existence of LD and its co-infections, and more effectively aid with developing a treatment plan. By testing for the genes of infectious organisms and quantifying the data, several advantages will accrue including: (1) better accuracy in telling if the patient has LD and plasmids, (2) better ability to fight LD, (3) better quality testing, (4) better information on quantity, (5) better selection of drug, and (6) discarding of non-drug targets for personalized treatment. It does not rely on detecting antibodies that may or may not be present. It helps determine a treatment plan that is best suited to attack the infections specific to the patient. It also allows tracking improvement of the patient's condition. Further, quantifiable data give hard evidence of the existence of CLD aiding in spreading awareness, hopefully producing a better future for those who suffer from this debilitating disease.

- **Next generation sequencing:** Whereas PCR testing can only detect predetermined large strands of DNA, next generation sequencing (NGS) is a newer technology that is capable of sequencing millions of small strands of DNA from a single blood sample. The test sensitivity would thus be potentially larger. A clinical trial currently conducted at Stony Brooks University, New York, will investigate the capability of NGS to detect *Borrelia burgdorferi* DNA in the blood of pediatric patients with LD at all suspected phases or stages of the disease. The test will begin before or up to 24 hours after the first dose of antibiotics is administered.

- **Neurologic tests of *neuroborreliosis* cases:** There is a distinction between Europe and North America. In Europe, *neuroborreliosis* is usually caused by *Borrelia garinii* and almost always involves lymphocytic pleocytosis (LPC). In LPC, the densities of lymphocytes (infection-fighting cells) and protein in the cerebrospinal fluid (CSF) typically rise to characteristically abnormal levels, while glucose levels remain normal. Additionally, the immune system produces antibodies against Lyme inside the intrathecal space, which contains the CSF. Demonstration by lumbar puncture, CSF analysis of pleocytosis, and intrathecal antibody production are required for definite diagnosis of *neuroborreliosis* in Europe - (except in cases of peripheral neuropathy associated with *acrodermatitis chronica atrophicans* (ACA) - which is usually caused by *Borrelia afzelli* and confirmed by blood antibody tests. In North America, *neuroborreliosis* is caused by *Borrelia burgdorferi*. It may not be accompanied by the same CSF signs. A negative diagnosis of central nervous system (CNS) *neuroborreliosis* does not exclude *neuroborreliosis*. American guidelines consider CSF analysis optional when symptoms appear to be confined to the peripheral nervous system (PNS), e.g. facial palsy without overt meningitis symptoms.

Those patients that present with neurological Lyme need specialized care that is critical to their improvement.

- **Cardiologic tests of Lyme carditis:** In Lyme carditis, electrocardiograms (EKG) are used to evidence heart conduction abnormalities while echocardiography may show myocardial dysfunction. Biopsy and confirmation of *Borrelia* cells in myocardial tissue may be used in specific cases but are usually not done because of the risks of the procedure.

An overall summary of the diagnostic methodology for late-disseminated LD is provided in Table 5.3.

Table 5.3 – Diagnostic methodology: Late-disseminated disease

Principles	Test	Advantages	Shortcomings
Blood tests	Positive AB	Can exclude LD as possible cause of observed symptoms	Misses diagnosis of: o Chronic fatigue syndrome o Crohn's disease o Fibromyalgia o HIV o Lupus o Multiple sclerosis o Rheumatoid arthritis o Other autoimmune and neurodegenerative diseases
	Brain pathogens: o Cytomegalovirus o Herpes o Other pathogens	o See Sidebar 5.1 o See Sidebar 5.2 o See Sidebar 5.2	
Encephalitis	o Familial o Autoimmune	See Sidebar 5.3	

Key: *AB: antibodies; LD: Lyme disease*

- **Single-photon emission computed tomography** (SPECT): SPECT images show numerous areas where an insufficient amount of blood is being delivered to the cortex and subcortical white matter. They can also identify abnormalities in the brain of a person affected with this disease. However, SPECT images are known to be nonspecific because they show a heterogeneous pattern in the imaging. The abnormalities seen in these images are very similar to those seen in people with cerebral vacuities and Creutzfeldt-Jakob disease (CJD), which makes them questionable.

Differentiating from confounding diseases

Lyme disease

Community clinics misdiagnose 23%-28% of EM rashes and 83% of other objective manifestations of early LD. EM rashes are often misdiagnosed as spider webs, cellulitis, or shingles. Many misdiagnoses are credited to the widespread misconception that EM rashes should look like a "bull's eye". Actually, the key distinguishing features of the EM rash are not its anatomical appearance but its characteristic mechanical features: (1) the speed and extent to which it expands, respectively up to 2–3 cm/day and a diameter of at least 5 cm, and in 50% of cases more than 16 cm; and (2) The rash expands away from the center, which may or may not look different or be separated by a ring-like clearing from the rest of the rash:

- **Spider webs:** Compared to EM rashes, spider bites are more common in the limbs, tend to be more painful and itchy or become swollen, and some may even cause necrosis (sinking dark blue patch of dead skin).

- **Cellulitis:** It most commonly develops around a wound or ulcer, is rarely circular, and is more likely to become swollen and tender. EM rashes often appear at tissue folds (armpit, groin, abdomen, back of knee) and other sites that are unusual for cellulitis.

- **Shingles:** Like Lyme, shingles often begins with headache, fever, and fatigue, which are followed by pain or numbness. However, unlike Lyme, in shingles, these symptoms are usually followed by the appearance of rashes composed of multiple small blisters along a nerve's dermatome. Shingles can also be confirmed by quick laboratory tests.

- **Lyme disease facial palsy (LDFP):** Facial palsy caused by Lyme disease (LDFP) is often misdiagnosed as Bell's palsy (BP). Although BP is the most common type of one-sided facial palsy (about 70% of cases), LDFP can account for only about 25% of cases of facial palsy in areas where LD is common. Compared to LDFP, BP much less frequently affects both sides of the face. Even though LDFP and BP have similar symptoms and evolve similarly if untreated, corticosteroid treatment is beneficial for BP while being detrimental for LDFP.

  The likelihood of LDFP should be based on recent history of exposure to a likely tick habitat during warmer months, EM rash, viral-like symptoms (headache, fever, and/or palsy in both sides of the face). If it is more than minimal, empiric therapy with antibiotics should be initiated, without corticosteroids, and reevaluated upon completion of laboratory tests for LD.

- **Lyme lymphocytic meningitis (LLM):** Unlike viral meningitis, LLM tends to not cause fever, last longer, and recur. It is also characterized by its possible co-occurrence with EM rash, FP, or partial vision obstruction and having much lower percentage of polymorphonuclear leukocytes (PNL) in CSF.

- **Lyme radiculopathy (LR):** Affecting the limbs, LR is often misdiagnosed as a radiculopathy caused by nerve root compression, such as sciatica. Most LR cases are compressive and resolve with conservative treatment (e.g., rest) within 4–6 weeks. Nonetheless, guidelines for managing LR recommend first evaluating risks of other possible causes that, although less frequent, require immediate diagnosis and treatment, including infections such as Lyme and shingles. A

Lyme disease

history of outdoor activities in likely tick habitats in the last 3 months, possibly followed by a rash or viral-like symptoms and current headache, other symptoms of lymphocytic meningitis, or FP would lead to suspicion of LD and recommendation of serological and lumbar puncture tests for confirmation. LR affecting the trunk can be misdiagnosed as a myriad of other conditions such as diverticulitis and coronary artery syndrome (CAS).

Against the EM rash, many of the several confounding diseases are summarized in Table 5.4.

Table 5.4 – Differentiating Lyme from confounding diseases

Disease	Appearance/ Characteristics	Confounding factor
Erythema migrans rash	o Not always a "bull's eye" o Diameter 5cm (in 50% of cases: up to 16cm) o Expands 2-3cm/day	May be present in other diseases
Spider web	o More common in the limbs o May cause necrosis	Painful, itchy, swollen
Cellulitis	Rarely circular	o Develops around a wound or ulcer o Appears at tissue folds that are rare for cellulitis
Shingles	Rashes composed of multiple small blisters along a nerve's dermatome	Begins with headache, fever, fatigue followed by pain or numbness
Facial palsy (often diagnosed as Bells' palsy)	Affects both face sides	Comparison of Bell's palsy to LDFP: o Similar symptoms o Similar evolution if untreated o Corticosteroids beneficial for Bell, not so much for LDFP
Lyme lymphocytic meningitis	o Different from viral meningitis o No fever, lasts longer, recurs o Possibly co-occurring with EM rash, facial palsy, or partial vision obstruction o Much lower percentage of people in CSF	Viral meningitis
Lyme radiculopathy	o Affects both limbs o Compressive o Resolves within 4-6 weeks with conservative treatment (rest)	o Ordinary radiculopathy caused by nerve root compression (e.g., sciatica) o Diverticulitis o Coronary artery syndrome

Key: *CAS: Coronary artery syndrome; CSF: Cerebrospinal fluid; LDFP: Lyme disease facial palsy; LLM: Lyme lymphocytic meningitis; PML: Polymorphonuclear leukocytes*

Take-home points

- It is important to differentiate the actual bacterial strain as there are a number of different strains including hybridized strains that may require different treatments. The importance of the particular bacterial strain that is responsible for the infection is only just being revealed with researchers observing patterns of LD symptoms associated with each type of *Borrelia* bacteria. Thus, knowing the prevalent strains of spirochete in an area can allow the population to be on the lookout for specific symptoms in order to catch LD in its localized early (acute) stage rather than as it disseminates and becomes harder to treat. Early LD symptoms can be easily overlooked and those who are aware of the risk of LD in their communities are usually more likely to seek early medical attention after observing more subtle symptoms of infection.

- The problem in current testing for Lyme disease is the high likelihood of receiving a false negative, which occurs when a test produces results indicating that a disease is not present when, in reality, it is.

- ELISA and Western blot are the standard testing methods used to diagnose Lyme disease. These older tests look for the presence of certain antibodies to produce a positive result when searching for Lyme disease. However, Lyme disease patients are often immune-compromised and their bodies may not be producing the anti-bodies necessary. In addition, these tests are not quantitative and, therefore, are unable to provide conclusive data.

- Lyme disease patients are often caught in a vicious cycle of immune depression that begins with the initial infection.

- When bitten by a tick with Lyme disease, co-infective agents that the tick may possess are also transferred. Once in the body, these several infections release a multitude of endotoxins, neurotoxins, and biotoxins that confuse the immune system and cause it to potentially attack its own cells resulting in autoimmune-like symptoms and chronic inflammation throughout the brain and body.

- The confusion caused by toxins, and other factors that depress the immune system, makes the patient's body more susceptible to opportunistic secondary fungal, viral, parasitic, and other bacterial infections which work to further impair the patient's immune system.

- With the immune system severely impaired by Lyme and its co-infections, the traditional diagnostic tests may not be accurate. Worse, co-infections, biotoxins, endotoxins, mycotoxins, neurotoxins, autoimmune attacks, and opportunistic secondary infections can produce a multitude of symptoms making it difficult to recognize Lyme.

- While most infections are tick-borne in nature, chronic Lyme disease also includes complicating primary and secondary co-infections that may also be present. It is the combination of these multiple infections with other complications that contribute to the patients' debilitating set of physical and neurological symptoms. Only in-depth, up-to-par laboratory testing can identify these several infections. Correctly identifying and vigorously treating all these infections is the key in producing lasting results against chronic Lyme disease.

Lyme disease

- Diagnosis is based upon a combination of symptoms, history of tick exposure, and possibly testing for specific antibodies in the blood. The common tests (Western Blot and ELISA) used to confirm a Lyme diagnosis can report incorrect results in 50% of the time.

- To complicate matters, Lyme patients with neurological symptoms are often misdiagnosed with one or more neurological diseases.

- There are four basic principles involved in the diagnosis of chronic Lyme disease: (1) History of possible exposure to infected ticks; (2) signs and symptoms observed; (3) objective physical findings (such as *erythema migrans* rash, facial palsy, arthritis, etc.); and possibly (4) laboratory tests.

- The several diagnostic tests have been abundantly discussed in cases of localized (or early Lyme) disease, early-disseminated, and late-disseminated Lyme. These include: ELISA, Western blot, dark-field microscopy, Ispot, polymerase chain reaction, and genomic testing. This latter test, in particular, is the most accurate of all but it requires further development and validation. In the latter instance, so-called next generation sequencing may allow greater sensitivity than the polymerase chain reaction test. Other neurologic and cardiologic tests have also been discussed.

- Though controversial, certain neuroimaging tests (magnetic resonance, single-photon emission computed tomography) can provide data that are diagnostically helpful.

- There are various confounding diseases that need to be differentiated against including spider webs, cellulitis, shingles, facial palsy, Lyme lymphocytic meningitis, Lyme radiculopathy. diverticulitis, and coronary artery syndrome.

- Late-stage Lyme disease may be misdiagnosed as multiple sclerosis, rheumatoid arthritis, fibromyalgia, chronic fatigue syndrome, lupus, Crohn's disease, HIV or other autoimmune and neurodegenerative diseases.

6

Clinical practice guidelines

Clinical practice guidelines (CPG) are often used as reference by physicians for LD treatment and treatment of other tick-borne diseases. Several CPL have been issued by governmental and professional organizations. The only CPG posted on the National Guidelines Clearinghouse (NGC), under the auspices of the (U.S.) Department of Health & Human Services (DHHS), are those adhering to newly revised National Academy of Medicine (NAM), formerly the Institute of Medicine (IOM), standards for guidelines: the International Lyme & Associated Diseases Society (ILADS) Lyme Guidelines, which address the usefulness of antibiotic prophylaxis for tick bite, the effectiveness of EM treatment, and the tole of antibiotics in the treatment of persistent LD symptoms.

The goals of medical care in LD should always be to prevent the illness whenever possible and to cure the illness when it occurs ("treating the treatable"). When this is not possible, the emphasis for treatment should be on reducing patient morbidity by reducing patient risks for developing the chronic form of the disease and on reducing the serious morbidity associated with these disease forms. Primary prevention is preferred by effectively treating a tick bite, secondary prevention by treating an EM rash sufficiently so as to restore health and prevent disease progression, and tertiary prevention by treating patients whose illness may be responsive to additional therapy, thereby reducing the morbidity associated with the chronic forms of the disease.

Because of their importance, I briefly review below the available practice guidelines.

The Infectious Diseases Society of America guidelines

The Infectious Diseases Society of America (IDSA) has updated its 2019 clinical practice guidelines for clinical infectious diseases in general. These are among high-risk populations. Unfortunately, the guidelines are too general and broadly address all infectious diseases, not specifically LD. Further, they are aimed at controlling outbreaks of infectious diseases in certain populations. They consider the care of children, pregnant and *postpartum* women, and non-pregnant adults, and include special considerations for patients who are severely immunocompromised such as hematopoietic stem cell and solid-organ transplant recipients. While the target audience includes primary care clinicians, obstetricians, emergency medicine providers, hospitalists, and infectious disease specialists, the guidelines may also be useful for occupational health physicians and clinicians working in long-term care facilities.

The guidelines add new information on diagnostic testing, use of antivirals, considerations of when to use antibiotics, and when to test for antiviral resistance. They also present evidence on harm associated with routine use of corticosteroids. The process followed that used in the development of previous IDSA guidelines that included a systematic weighting of the strength of recommendations and quality of evidence based upon the (U.S.) Public Health Service (PHS) grading system for ranking recommendations in clinical guidelines. Unfortunately, the recommendations exclusively address seasonal influenza which, although undeniably important, is a different infectious disease than LD.

The Association of State and Territorial Public Health Laboratory Directors guidelines

In 1994, the (U.S.) Association of State and Territorial Public Health Laboratory Directors (ASTPHD), the CDC&P, the (U.S.) Food & Drug Administration (FDA), the (U.S.) National Institutes of Health (NIH), the (U.S.) Council of State and Territorial Epidemiologists (CSTE), and the National Committee for Clinical Laboratory Standards (NCCLS) convened the Second National Conference on Serologic Diagnosis of Lyme Disease during which they recommended a two-test methodology using a sensitive enzyme immunoassay (EIA) or immunofluorescence assay (IFA) as a first test, followed by a western immunoblot assay (IBA) for specimens yielding positive or equivocal results.

Regarding future tests, the report advised that *"...evaluation of new serologic assays include blind testing against a comprehensive challenge panel, and that new assays should only be recommended if their specificity, sensitivity, and precision equaled or surpassed the performance of tests used in the recommended two-test procedure"*.

With support from NIH and to assist serologic test developers, CDC&P made available a comprehensive panel of sera from patients with various stages of LD and other conditions, as well as healthy persons. Thus, on 7/2/19, the FDA cleared several LD serologic assays with new indications for use based on a modified two-test methodology. This modified methodology uses a second EIA in place of a Western immunoblot assay. Clearance by FDA of the new LD assays indicates that test performance has been evaluated and is "substantially equivalent to or better than" a legally marketed predicate test.

Unfortunately, even though updated, the basis for the FDA/CDC&P recommendations rely too much on old technology, do not account for newer technological developments, and do not espouse principles of modern integrative and personalized medicine. Still further, conventional testing often provides false negatives when diagnosing Lyme *borreliosis*, resulting in a monumental failure to provide the needed critical early detection and treatment. I have discussed these newer technological advances and the associated tests in Chapter 4.

The (U.S.) CDC&P original guidelines for serologic analyses

For LD

When assessing a LD patient, health care providers should consider:

- The signs & symptoms of LD;
- The likelihood that the patient has been exposed to infected black-legged ticks;
- The possibility that other illnesses may cause similar symptoms; and
- Results of laboratory tests, when indicated.

LD is a tick-borne zoonosis for which serologic testing is currently the principal means of laboratory diagnosis. The diagnosis algorithm recommended by CDC&P consists of a two-step testing process that can be done using the same blood sample pending the development of new tests as alternatives to one or both steps. If the first step is negative, no further testing is recommended. On the other hand, if it is positive or indeterminate (sometimes called "equivocal"), the second step should be performed. The overall result is positive only when the first test is positive (or equivocal) and the second test is positive (or for some tests equivocal). (Figure 6.1)

Figure 6.1 – CDC&P guideline for Lyme disease serology

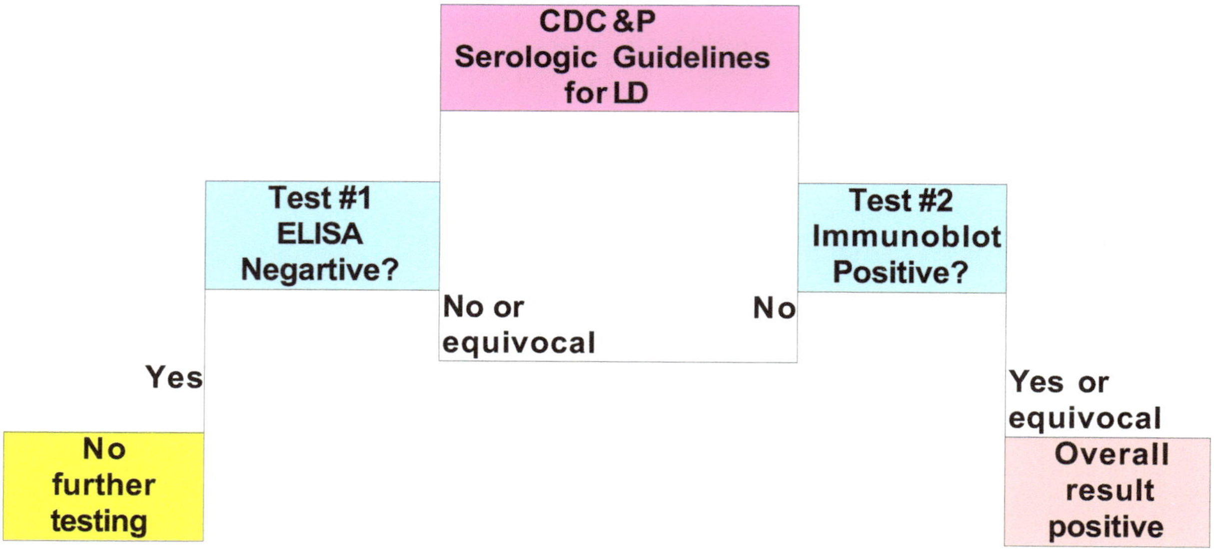

In so doing, key points to remember are the following:

- Most LD tests are designed to detect antibodies made by the body in response to infection;
- Antibodies can take several weeks to develop, so patients may test negative if infected only recently;

- Antibodies normally persist in the blood for months or even years after the infection is gone; therefore, the test cannot be used to determine cure; and

- Infection with other diseases, including some tick-borne diseases, or some viral, bacterial, or autoimmune diseases, can result in false positive test results.

The above methodology was jointly recommended by the CDC&P together with the ASTPHLD, the FDA, the NIH, the CSTE, and the NCCLS at their Conference SNCSDLD held 27-29 October 1994.

Regarding serologic test performance and interpretation, the two-test approach for active disease and for previous infection using ELISA (a sensitive enzyme immunoassay) or IFA followed by a Western (immuno)blot was the algorithm of choice (see my discussion in Chapter 4). Specifically:

- All specimens positive or equivocal by a sensitive ELISA or IFA should be tested by a standardized Western immunoblot. Specimens negative by a sensitive ELISA or IFA need not be tested further.

- When Western immunoblot is used during the first 4 weeks of disease onset (early LD), both immunoglobulin M (IgM) and immunoglobulin G (IgG) procedures should be performed.

- A positive IgM test result alone is not recommended for use in determining active disease in persons with illness greater than 1 month's duration because the likelihood of a false-positive test result for a current infection is high for these persons. An IgM immunoblot is considered positive if two of the following three bands are present: 24 kDa (OspC)*, 39 kDa (BmpA), and 41 kDa (Fla).

- If a patient with suspected early LD has a negative serology, serologic evidence of infection is best obtained by the testing of paired acute- and convalescent-phase serum samples. Serum samples from persons with disseminated or late-stage LD almost always have a strong IgG response to *Borrellia burgdorferi* antigens. An IgG immunoblot should be considered positive if five of the following 10 bands are present: 18 kDa, 21 kDa (OspC), 28 kDa, 30 kDa, 39 kDa (BmpA), 41 kDa (Fla), 45 kDa, 58 kDa (not GroEL), 66 kDa, and 93 kDa (2).

For CLD

Recently, CDC&P released a case study regarding the treatment of CLD based on a few published studies involving a small number of patients. It concluded that antibiotics do not provide long-term benefits but rather gave rise to other complicating infections. Based on these outcomes, it also inferred, that patients should not be so treated but rather referred to other specialists (rheumatologists, psychiatrists, pain management specialists, and neurologists). However, while on the surface, such referrals appear sound and reasonable, they do not eliminate the root cause of the disease, which is the infections. Unfortunately, these symptomatic treatments ultimately leave patients in a life-time of suffering and in a condition that only worsens over time. However, although providing temporary relief, IV antibiotics were never effective for numerous reasons. The important question and further case study work should help answer is why IV antibiotics alone do not work.

The updated FDA/CDC&P guidelines for serologic analyses

Very recently (16 August 2019), CDC&P has updated its recommendations for serologic diagnosis of LD. As is known, serologic testing is the principal means of laboratory diagnosis of LD. The current recommendations have been summarized in the previous section. They have now been updated following FDA clearance (on 29 July 2019) of several LD serologic assays with new indications for use, allowing for an enzyme immunoassay (EIA) rather than a a Western immunoblot assay (IBA) as the second test in the LD testing algorithm.

The CDC&P recommendations for health care providers

When a patient seeks care after a tick bite, topics to discuss should include:

- Tick removal (if still present), degree of engorgement, and identification;

- LD prophylaxis, as determined by the tick species and degree of engorgement; and

- Symptom watch.

Antimicrobial prophylaxis for the prevention of LD

Antimicrobial prophylaxis for the prevention of LD following a tick bite can be started within 72 hours of tick removal. It may be beneficial in certain circumstances. A single dose of the antibiotic doxycycline can lower the risk of LD when:

- The tick bite occurred in a State where LD incidence is high (Maryland, Massachusetts, Minnesota, Missouri, New York, Pennsylvania, Virginia, West Virginia, and Wisconsin) or in an area where >20% of ticks are infected with *Borrelia burgdorferi,* and the patient has no contraindication to doxycyclin. The local health department can usually provide information about tick infection rates in its area;

- If a person is suspected of acute tick-borne disease, including early LD or Rocky Mountain spotted fever, treatment should be initiated as soon as possible, rather than waiting for laboratory results, which may be insensitive in early illness;

- Location of tick exposure can guide the differential diagnosis;

- The attached tick can be identified as an adult or nymphal black-legged tick;

- The estimated time of attachment is ≥36 hours based on the degree of tick engorgement with blood or likely time-of-exposure to the tick; and

- The patient has no contraindication to doxycycline.

Prophylaxis can be started within 72 hours of tick removal (see Table 6.1). It is to be noted that antibiotic treatment following a tick bite is not recommended as a means to prevent tick-borne diseases other than LD (such as *anaplasmosis, babesiosis, ehrlichiosis*, and Rocky Mountain spotted fever). There is no evidence this practice is effective, and it may simply delay the onset of disease.

Table 6.1 - Recommended Lyme disease post-exposure prophylaxis

Age category	Drug	Dosage	Maximum	duration
Adults	Doxycycline	200 mg orally	N/A	Once
Children (weighting less than 45kg)	Doxycycline	4.4 mg/kg orally	200 mg	Once (Note: ILADS is against this use)

ILADS: International Lyme and Associated Diseases Society

Symptom watch

The patient should be encouraged to watch for fever, rash, or flu-like illness in the weeks after a tick bite.

Laboratory tests that are *not* recommended

Some laboratories offer LD testing using assays whose accuracy and clinical usefulness have not been adequately established. Examples of invalidated tests include:

- Capture assays for antigens in urine;

- Culture, immunofluorescence staining, or cell sorting of cell wall-deficient or cystic forms of *Borrelia burgdorferi*;

- Lymphocyte transformation tests;

- Quantitative CD57 lymphocyte assays;

- "Reverse Western blots";

- In-house criteria for interpretation of immunoblots;

- Measurements of antibodies in joint fluid (synovial fluid); and

- IgM or IgG tests without a previous ELISA/EIA/IFA test.

Lyme disease

Take-home points

- The Infectious Diseases Society of America has updated its 2019 clinical practice guidelines for clinical infectious diseases, in general. Unfortunately, they are too general and broadly address all infectious diseases, not specifically Lyme Disease. Further, they are aimed at controlling outbreaks of infectious diseases in certain (high-risk) populations.

- The guidelines add new information on diagnostic testing, use of antivirals, considerations of when to use antibiotics, when to test for antiviral resistance, and present evidence on harm associated with routine use of corticosteroids.

- In 1994, several U.S. governmental entities - the Association of State and Territorial Public Health Laboratory Directors, the Centers for Disease Control & Prevention, the National Institutes of Health, the Council of State and Territorial Epidemiologists, and the National Committee for Clinical Laboratory Standards - convened the Second National Conference on Serologic Diagnosis of Lyme Disease during which they recommended a two-test methodology using a sensitive enzyme immunoassay (EIA) or immunofluorescence assay (IFA) as a first test, followed by a western immunoblot assay (IBA) for specimens yielding positive or equivocal results.

- Unfortunately, even though updated, the basis for the FDA/CDC&P recommendations rely too much on old technology, do not account for newer technological developments (such as genomic testing), and do not espouse principles of modern integrative and personalized medicine. Still further, conventional testing often provides false negatives when diagnosing Lyme *borreliosis*, resulting in a monumental failure to provide the needed critical early detection and treatment.

- The CDC&P original guidelines for serologic analyses recommend that, when assessing a Lyme disease patient, health care providers should consider: the signs & symptoms of the disease, the likelihood that the patient has been exposed to infected black-legged ticks, the possibility that other illnesses may cause similar symptoms, and results of laboratory tests (when indicated).

- Most Lyme disease tests are designed to detect antibodies made by the body in response to infection. However, antibodies can take several weeks to develop so patients may test negative if infected only recently. Further, antibodies normally persist in the blood for months or even years after the infection is gone; therefore, the test cannot be used to determine cure. Further, infection with other diseases, including some tick-borne diseases, or some viral, bacterial, or autoimmune diseases, can result in false positive test results.

- All specimens positive or equivocal by a sensitive ELISA or IFA should be tested by a standardized Western immunoblot. Specimens negative by a sensitive ELISA or IFA need not be tested further.

Lyme disease

- When Western immunoblot is used during the first 4 weeks of disease onset, both immunoglobulin M (IgM) and immunoglobulin G (IgG) procedures should be performed.

- Very recently (16 August 2019), the CDC&P have updated its recommendations for the serologic diagnosis of Lyme disease with new indications for use, allowing for an enzyme immunoassay (EIA) rather than a Western immunoblot assay (IBA) as the second test in the LD testing algorithm.

- The CDC&P have also provided recommendations for health care providers when a patient seeks care after a tick bite, including: tick removal, degree of engorgement, and identification; prophylaxis, as determined by the tick species and degree of engorgement; and symptom watch.

- Antimicrobial prophylaxis for the prevention of Lyme disease following a tick bite can be started within 72 hours of tick removal. It may be beneficial in certain circumstances. A single dose of the antibiotic doxycycline can lower the risk of the disease when: the tick bite occurred in a State where Lyme disease incidence is high or in an area where >20% of ticks are infected with *Borrelia burgdorferi,* and the patient has no contraindication to doxycyclin. The local health department can usually provide information about tick infection rates in the area under its jurisdiction.

- If a person is suspected of acute tick-borne disease, treatment should be initiated as soon as possible, rather than waiting for laboratory results, which may be insensitive in early illness.

- Prophylaxis can be started within 72 hours of tick removal. It is to be noted that antibiotic treatment following a tick bite is not recommended as a means to prevent tick-borne diseases other than LD (such as *anaplasmosis, babesiosis, ehrlichiosis*, and Rocky Mountain spotted fever). There is no evidence that this practice is effective, and it may simply delay the onset of disease.

The interested reader will find in Sidebar 6.1 a brief discussion of the evidence-based guidelines that have been developed for the management of Lyme disease patients. Sidebar 6.2 recaps the management and treatment guidelines developed by the International Lyme and Associated Diseases Society (ILADS).

Sidebar 6.1
Evidence-based guidelines for the management of LD patients

Evidence-based guidelines for the management of LD patients were developed by the International Lyme and Associated Diseases Society (ILADS) to replace its earlier 2004 guidelines. They address three clinical questions, namely, the:

- ***Usefulness of antibiotic prophylaxis:*** for known tick bites;

- *Effectiveness of erythema migrans treatment (EM)*; and

- *Role of antibiotic re-treatment in patients with persistent manifestations of LD*.

Although the intended users of the new ILADS guidelines are healthcare providers who evaluate and manage patients with LD, the guidelines could also be of interest to patients themselves for their greater understanding of their condition, the appropriateness of the treatment(s) they are following, and (hopefully) the progress they are making. This would contribute to informing and empowering patients to engage in shared decision-making. These clinical practice guidelines (CPG) are intended to assist clinicians by presenting evidence-based treatment recommendations, which follow the Grading of Recommendations Assessment, Development, and Evaluation (GRADE) system. The GRADE scheme itself is a continually evolving system. The guidelines attempt to incorporate the current state of GRADE. They ensure a transparent and trustworthy guideline process. They are not, however, intended to be the sole source of guidance in managing LD and should neither be viewed as a substitute for clinical judgment nor used to establish treatment protocols.

But,... what is evidence-based medicine?

Evidence-based medicine is "the integration of the best available research evidence with clinical expertise and patient values". ILADS anticipates performing GRADE assessments on additional topics related to the diagnosis and treatment of tick-borne diseases in the future.

For the reader's information, the GRADE scheme classifies the quality of the evidence as "high", "moderate", "low", or "very low". The quality of evidence is thus:

- **From randomized controlled trials (RCTs):** Initially rated as high, it may be downgraded based on five limitations: (a) study bias, (b) publication bias, (c) indirectness (generalizability), (d) imprecision, and (e) inconsistency;

- **From observational studies:** Generally low, but may be upgraded based on a large effect or dose–response gradient (DRG).

Rather than labeling recommendations as strong or weak, these guidelines use the terms "recommendation" or "strong recommendation" for or against a medical intervention.

What is the evidence quality for LD?

Although LD is not rare, its treatment has not attracted pharmaceutical interest (see Chapter 9 on vaccines for LD). The evidence base for treating LD is best described as sparse, conflicting, and emerging. The evidence quality was assessed as "very low". This is consistent with the evidence base for the infectious field as a whole. Indeed, the majority of recommendations in infectious disease medicine generally are based on low-quality evidence.

Sidebar 6.2

ILADS management and treatment guidelines

The International Lyme and Associated Diseases Society (ILADS) has issued management and treatment guidelines, which I am providing here in an attempt to balance the information available to patients for their medial care.

Management guidelines

Evidence-based guidelines for the management of patients with Lyme disease were developed by ILADS. The guidelines address three clinical questions:

- *Usefulness of antibiotic prophylaxis for known tick bites;*
- *Effectiveness of erythema migrans (EM) treatment;* and
- *Role of antibiotic re-treatment in patients with persistent manifestations of LD.*

These clinical practice guidelines are intended to assist clinicians by presenting evidence-based treatment recommendations. They are not intended to be the sole source of guidance in managing LD and they should not be viewed as a substitute for clinical judgment nor used to establish treatment protocols.

Treatment guidelines

The optimal treatment regimen for the management of known tick bites, erythema *migrans* (EM) rashes, and persistent disease has not yet been determined. Accordingly, it is too early to standardize restrictive protocols. Nonetheless, ILADS does make recommendations for each of these clinical situations:

- *Against the use of a single 200 mg dose of doxycycline for the prevention of LD:* Not only is it unlikely to be highly efficacious, failed therapy in a human trial led to a seronegative disease state;

- *20-day doxycycline treatment for known black-legged tick bites:* (barring any contra-indications). This is based on animal studies;

- *4-6 weeks of antibiotic treatment days for EM rashes:* Antibiotics include doxycycline, amoxicillin or cefuroxime. A minimum of 21 days of azithromycin is also acceptable, especially in Europe. All patients should be reassessed at the end of their initial therapy and, when necessary, antibiotic therapy should be extended.

- *Evaluation for other potential causes before instituting additional antibiotic therapy:* For patients with persistent symptoms and signs of LD.

- *Antibiotic re-treatment:* when a chronic Lyme infection is judged to be a possible cause of the ongoing manifestations and the patient has an impaired quality of life.

Clinical judgment and shared decision-making

Given the number of clinical variables that must be managed and the heterogeneity within the patient population, clinical judgment is crucial to the provision of patient-centered care. Patient goals and values regarding treatment options must be identified and strongly considered during a shared decision-making process.

Reconciling divergent guidelines

Conflicting guidelines most often result when evidence is weak; when developers differ in their underlying values, approach to evidence reviews, synthesis or interpretation; and/or when developers have varying assumptions about intervention benefits and harms. These should be reconciled.

7

Can LD be treated...and how?

Treatment varies depending on the stage of the disease as I will now review and discuss.

Localized (or early) LD

For early LD, a short course of oral antibiotics is curative in the majority of the cases. People treated with appropriate antibiotics in the early stages of LD usually recover rapidly and completely.

Antibiotics commonly used for oral treatment include amoxicillin, cefuroxime axetil, or doxycycline, The treatment regimens last 2-3 weeks. In more complicated cases, the disease can usually be successfully treated with three to four weeks of antibiotic therapy. In the case of doxycycline, recent publications suggest greater efficacy for shorter courses of treatment (Table 7.1).

Treatment regimens listed in the following Table are for localized (or early) LD. These regimens are guidelines only and may need to be adjusted depending on a person's age, weight, medical history, underlying health conditions, pregnancy status, or allergies.

Table 7.1 – Treatment regimens for localized Lyme disease
(adapted from CDC&P)

Age category	Drug dosage	Dosage (orally)	Maximum	Duration (days)
Adults	Amoxicillin	500 mg (3 times a day)	N/A	14-21
	Cefuroxime acetyl	500 mg (2 times a day)	N/A	14-21
	Doxyxycline	100 mg (2 times a day)	N/A	10-21
Children	Amoxicillin	50 mg/kg per day divided into 3 doses	500 mg/dose	14-21
	Cefuroxime acetyl	30 mg/kg per day divided into 2 doses	500 mg/dose	14-21

	Doxyxycline	4 mg/kg per day divided into 2 doses	100 mg/dose	10-21

Note that for people intolerant of amoxicillin, cefuroxime axetil, and doxycycline, the macrolides azithromycin, clarithromycin, or erythromycin may be used, although they have a lower efficacy. People treated with macrolides should be closely monitored to ensure that symptoms resolve. Also, people with certain neurological or cardiac forms of illness may require intravenous treatment with antibiotics such as ceftriaxone or penicillin (Table 7.2).

Table 7.2 – Regimens for people with antibiotic intolerance or neurological or cardiac illnesses

Medical condition	Alternative treatment
Intolerance of traditional antibiotics (amoxicilin, cefuroxime acetyl, or doxyxycline)	Macrolides (azithromycin, clarithromycin, or erythromycin)
Neurological or cardiac forms of illness	Other antibiotics (ceftriaxone, or penicillin)

Early-disseminated LD

Standard treatment typically lasts only four to six weeks, with extensive treatment widely believed to be unwarranted.

Late-disseminated LD

When residual *Borrelia* reemerges, patients often relapse with LD or more specifically with what is known as chronic Lyme disease (CLD) or, more commonly, Lyme disease complex (LDC). To further complicate matters, most patients diagnosed with LD are unaware or/and ignorant of the full "complex" of co-infections and neurotoxins that reside in their system. The simultaneous presence of multiple different infections in the body seriously complicates any potential treatment and eliminates the possibility of prescription medication as a viable, singular LD treatment.

Nonetheless, the improvement of the patient's overall symptoms held over time is really the best measure and final indicator of successful CLD treatment.

Post-treatment Lyme disease syndrome

Although most cases of LD can be cured with a 2-4 -week course of oral antibiotics, patients can sometimes have symptoms of pain, fatigue, joint and muscle aches, or difficulty thinking that last for more than 6 months after they finish treatment. This condition, discussed earlier, is termed "Post-Treatment Lyme Disease Syndrome (PLDS)". Why some patients experience PLDS is not known. There are three schools of thought that have tried to explain the situation:

- **Autoimmunity:** Some experts believe that *Borrelia burgdorferi* can trigger an autoimmune response causing symptoms that last well after the infection itself is gone. This is not an unusual happenstance as autoimmune responses are known to occur following other infections, including campylobacter (Guillain-Barré syndrome, GBS), chlamydia (Reiter's syndrome, RS), and strep throat (rheumatic heart disease);

- **Other, persistent non-LD infection:** Other experts hypothesize that PLDS results from a persistent but difficult to detect infection; and

- **Other unrelated causes:** Still others believe that the symptoms of PLDS are due to other causes unrelated to the patient's *Borrelia burgdorferi* infection.

Patients with PLDS usually get better over time, but it can take many months to feel completely well. Nonetheless, additional options for managing symptoms may be available. Again, long-term antibiotic treatment for ongoing symptoms associated with LD can entail possibly serious risks, accompanied by sometimes deadly complications.

There are no long-term benefits of long-term antibiotic therapy in PLDS

After being treated for LD, patients with PLDS have non-specific symptoms and no evidence of active infection. Unfortunately, there is no proven treatment for PLDS. Studies funded by the (U.S,) National Institutes of Health (NIH) have found that long-term outcomes are no better for patients who received additional prolonged antibiotic treatment than for patients who received placebo. Worse, long-term antibiotic treatment for PLDS was found not helpful, has been associated with serious, sometimes deadly complications, and can be dangerous.

The (U.S.) National Institute for Allergy & Infectious Diseases (NIAID) has looked at the potential benefits of long-term antibiotic therapy, funding three placebo-controlled clinical trials on the efficacy of this prolonged antibiotic therapy. Another trial was conducted in The Netherlands. I summarize below the findings of these four trials.

These trials were designed to ensure that several key parameters were addressed:

- Susceptibility: of *Borrelia burgdorferi* to the antibiotics used;

- Ability to crossing the blood-brain barrier, access the central nervous system, and persistence at effective levels throughout the course of therapy;

- Ability to kill bacteria living both outside and inside mammalian cells; and

- Safety and welfare of patients enrolled in the trials.

FIRST CLINICAL TRIAL:

The first clinical trial included two multi-center studies in patients with a well-documented history of previous LD but who reported symptoms common among people reporting PLDS (persistent pain, fatigue, impaired cognitive function, or unexplained numbness). Patients were treated with 30 days of an intravenous (IV) antibiotic followed by 60 days of an oral antibiotic.

While these studies reinforced the evidence that patients reporting PLDS symptoms have a severe impairment in overall physical health and quality of life, they provided no evidence of benefit from prolonged antibiotic therapy when compared with placebo.

SECOND CLINICAL TRIAL: (Results published in 2003)

Researchers examined the effect of 28 days of IV antibiotic compared with placebo in 55 patients reporting persistent, severe fatigue at least six months following treatment for laboratory-diagnosed LD. Patients were assessed for improvements in self-reported fatigue and cognitive function. The study yielded two results:

People receiving antibiotics did report a greater improvement in *fatigue* than those on placebo. However, there was no benefit to cognitive function.

Further, six of the study participants had serious adverse events associated with IV antibiotic use, four requiring hospitalization. Overall, the study authors concluded that additional antibiotic therapy for PLDS was not supported by the evidence.

THIRD CLINICAL TRIAL:

In this study, the researchers compared clinical improvement following 10 weeks of IV ceftriaxone *versus* IV placebo. The patients presented with objective memory impairment tests and were treated for LD. In a complicated statistical model, the ceftriaxone group showed a slightly greater improvement at 12 weeks, but at 24 weeks both the ceftriaxone and the placebo groups had improved similarly from baseline. In addition, adverse effects attributed to IV ceftriaxone occurred in 26% of patients. The authors concluded that because of the limited duration of the cognitive improvement and the risks involved, long-term antibiotic use for LD is not an effective strategy for cognitive improvement. More durable and safer treatment strategies are still needed.

FOURTH CLINICAL TRIAL: (conducted in the Netherlands in 2016)

The published results were subjected to rigorous statistical, editorial, and scientific peer review. It was again concluded that in patients with persistent symptoms attributed to LD, longer term treatment with antibiotics did not provide additional benefits compared with shorter term regimens.

The following legitimate questions could be raised following the above results:

Question # 1: If long-term antibiotic therapy is not effective, why do some people report improved symptoms following such treatment? Carefully designed, placebo-controlled studies have failed to demonstrate that prolonged antibiotic therapy is beneficial. Although isolated success stories are possible, such reports alone are not sufficient grounds to support a therapeutic approach. Here, it may well be that a positive response to prolonged antibiotic therapy may be due to the placebo effect, which was reported as high as 40% in the studies described above.

Question # 2: Does infection persist after antibiotic therapy? Several recent studies in nonhuman primates have suggested that *Borrelia burgdorferi* may persist in *animals* after antibiotic therapy. Thus:

Study # 1: Remnants of the bacterium remained in mice;

Study # 2: The intact bacterium persisted. (Note: However, it was not possible to culture these bacteria and it is not clear whether they are infectious.);

Study # 3: It replicated the earlier finding of persisting DNA but non-cultivatable bacterium using a mouse model; and

Study # 4: In 2017, persistent and metabolically active *Borrelia burgdorferi* was evidenced in rhesus macaques.

In light of these results, it is clear that additional research is needed to learn more about persistent infection in cell culture and animal models, and its potential implication for human disease.

Antibiotics are not the best option for CLD

Conventional Lyme treatment includes weeks or months of antibiotic therapy. This approach not only is proving to be ineffective, it is also potentially harmful with side effects that may include intestinal bleeding, blood clots in the lungs, and anemia. The longer the drugs are taken, the greater the risk of harm. At the least, the antibiotics disrupt the health of the gut bacteria or microbiota, impairing the immune function since at least 70% of the immune system cells reside in the gut. Further, they do not help the joints pain or nerve problems of the infection.

Some people think that the length of time antibiotics are taken makes the difference in their effectiveness. Results are proving otherwise. In two studies of CLD patients, three months of antibiotics were no better than a placebo. In some cases, the drugs may produce short-term improvement in the ailments but the improvement ceases once the course of antibiotics is finished. So, how does *Borrelia burgdorferi* escape such intense treatment? The answer lies in their incredible defenses including the following seven mechanisms:

Mechanism # 1: The Lyme bacteria can change their physical shape, "curling-up" into a ball so antibiotics cannot get into their system and kill them. This cyst form is resilient, and antibiotics are useless against it. Once the threat subsides, the microbes can return to their typical spiral shape;

Mechanism # 2: The Lyme bacteria are able to "talk" to each other. As soon as they detect antibiotics, they send out a distress call to the others. They can wind into cysts before the drugs harm them;

Mechanism # 3: The bacteria can "camouflage" themselves. The immune system notes a microbe's identity by memorizing its protein sequence or genetics. The microbes can alter little parts of their DNA continually changing their appearance, so they do not fit the code. The immune system then has to search for many codes, not one;

Mechanism # 4: The bacteria can "morph" their DNA every time antibiotics are taken. This makes them increasingly resistant to the drugs;

Mechanism # 5: The bacteria shed endotoxins from their cell wall when they die. Antibiotics cannot eliminate these harmful, inflammatory byproducts. Further, antibiotics cannot fix the heightened inflammation;

Mechanism # 6: The bacteria love to hide inside parasites. Just like us, parasites also have a microbiota. While we may have a mix of good and bad microbes in our gut, parasites are a Pandora's box of terrible microbes — one being Lyme. Even if we get rid of Lyme in the rest of our body, the bacteria hiding in parasites can reinfect us. This is why the disease may come back, despite long and intense antibiotic treatment; and

Mechanism # 7: To minimize harm from endotoxins and inflammation, the liver and kidneys need support, which antibiotics cannot provide.

Tests for the objective monitoring of treatment response

Unlike blood and intrathecal antibody tests, cerebrospinal fluid (CSF) pleocytosis tests revert to normal after infection ends. They can, therefore, be used as objective markers of treatment success and inform decisions on whether to retreat.

Also, in infection involving the peripheral nervous system (PNS), electromyography and nerve conduction studies can be used to monitor objectively the response to treatment.

Separately and independently, genomic testing may greatly help in prescribing the appropriate personalized treatment and accurately track its progress for each patient.

Take-home points

- Lyme disease treatment will vary depending on the stage of the infection (localized, early-disseminated, or late-disseminated).

- People treated with appropriate antibiotics in the early stages of LD usually recover rapidly and completely. Antibiotics commonly used for oral treatment include amoxicillin, cefuroxime axetil, or doxycycline, The treatment regimens last 2-3 weeks although, in the case of doxycycline, recent publications suggest greater efficacy for shorter courses.

- Treatments need to be adjusted depending on a person's age, weight, medical history, underlying health conditions, pregnancy status, or allergies.

Lyme disease

- For people intolerant of amoxicillin, cefuroxime axetil, and doxycycline, the macrolides azithromycin, clarithromycin, or erythromycin may be used, although they have a lower efficacy and require that patients be closely monitored to ensure that symptoms resolve. Also, people with certain neurological or cardiac forms of illness may require intravenous treatment with antibiotics such as ceftriaxone or penicillin.

- For early-disseminated Lyme disease, standard treatment typically lasts only four to six weeks, with extensive treatment widely believed to be unwarranted.

- For late-disseminated Lyme disease, when patients often relapse (what is known as chronic Lyme disease or Lyme disease complex), the simultaneous presence of multiple different infections in the body seriously complicates any potential treatment and eliminates the possibility of prescription medication as a viable, singular treatment. Improvement of the patient's overall symptoms held over time is really the best measure and final indicator of successful treatment.

- Although most cases of Lyme disease can be cured with a 2- to 4-week course of oral antibiotics, patients can sometimes have symptoms of pain, fatigue, or difficulty thinking that last for more than 6 months after they finish treatment, a condition called Post-Treatment Lyme Disease Syndrome.

- Why some patients experience post-treatment syndrome is not known. This may be caused by one or a combination of autoimmunity, other persistent non-Lyme infection, or even other unrelated causes.

- There is no proven treatment for post-treatment Lyme disease syndrome. Long-term outcomes are no better for patients who received additional prolonged antibiotic treatment than for patients who received placebo. Patients with the syndrome usually get better over time, but it can take many months to feel completely well.

- Patients with PLDS have non-specific symptoms after being treated for LD and have no evidence of active infection. For them, studies have shown that more antibiotic therapy is not helpful and can be dangerous.

- Long-term antibiotic treatment has been associated with serious, sometimes deadly complications.

- Through several placebo-controlled clinical trials, the (U.S.) National Institute for Allergy & Infectious diseases has confirmed that, compared to shorter-term antibiotic treatment, there is no evidence of benefit from prolonged antibiotic therapy when compared with placebo. Further, it is not an effective strategy for cognitive improvement.

- Several recent studies have suggested that the infection may persist in *animals* (mice, nonhuman primates, macaques) after antibiotic therapy. Additional research is clearly needed to learn more about persistent infection in cell culture and animal models and, especially, its potential implication for human disease.

- Antibiotics are not the best option for CLD because of the incredible defenses presented by *Borrelia burgdorferi.* consisting of seven mechanisms. The Lyme bacteria can: (a) change their physical shape, "curling-up" into a resistant cyst-form ball that antibiotics cannot penetrate; (b) "talk" to each other so that, as soon as one or some detect antibiotics, they send out a distress call to the others to wind into cysts before the drugs harm them; (c) can "camouflage" themselves so as not to be recognized by the immune system; (d) "morph" their DNA every time antibiotics are taken, making them increasingly resistant to the drugs; (e) "shed" harmful, inflammatory byproducts (endotoxins) from their cell wall that antibiotics cannot eliminate; (f) "hide" inside parasites; and (g) "reinfect" the body despite long and intense antibiotic treatment.

- Tests for the objective monitoring of treatment response are available in two instances: (a) cerebrospinal fluid pleocytosis and (b) electromyography and nerve conduction studies in infections involving the peripheral nervous system.

- Separately and independently, genomic testing may greatly help in prescribing the appropriate personalized treatment and accurately track its progress for each patient.

8

Is there a homeopathic treatment of Lyme and its co-infections?

How to beat CLD?

If antibiotics are not the best CLD treatment, what is? The answer is to work with your body and support the systems that have been overwhelmed. Treatment must address all the contributors to the disease in the right order. By supporting them first, the detoxifying organs will not be overwhelmed when the bacteria start to die. The recommended sequence is:

- Draining pathways;

- Detoxifying organs and the lymphatic system;

- Parasites; and

- Lyme.

Lyme is last on the list, *not* first. Going after it first will only put a heavier burden on the exhausted organs, which need to work well so they can help in the battle. The colon is the last stop in the body's detoxification system. If bowels are not moved at least 2–3 times a day, every process further up the line will be in jeopardy. Supporting the colon with intestinal moving herbs is essential as it opens the door for the next step. Once the colon is moving, the liver and kidneys need support. They filter out endotoxins from the dead pathogens. The lymphatic system also needs support, as the bacteria love to live there.

With the described support in place, the parasites that are harboring *Borrelia burgdorferi* can be tackled. Some researchers have recommended taking parasite-fighting herbs to knock-out this sneaky hiding place of the bacteria. Finally, one is ready to eliminate CLD.

I review below this plant "cocktail" treatment.

The plant "cocktail" treatment

As we know by now, CLD can weaken the immune system, ignite inflammation, squelch energy, provoke pain, and trigger brain fog. It can also generate harmful free radicals, disrupt mitochondria, and overwhelm detoxification pathways. No less than 21 different botanicals have been identified (there may be more), each one offering unique properties and several of them allegedly acting synergistically to help:

- Lower inflammation;
- Reduce joint pain;
- Fight free radical damage;
- Break up biofilm;
- Decrease viral load;
- Regulate the immune system;
- Purge parasites;
- Support detoxification; and
- Combat Lyme bacteria to beat CLD.

Table 8.1 provides an overview of these botanicals, indicating their compounds, alleged indication, alleged activity, and other issues. Claimed effects on Lyme and associated co-infections are highlighted. Other alleged benefits for other diseases (such as, for example, malaria) are also indicated for completeness.

IMPORTANT NOTE: In addition to the disclaimer on page 4 of this book, I emphasize that I have not researched these botanicals with regard to their alleged indication, activity, and other uses. Further, I have not ascertained whether they might be considered by the (U.S.) Food & Drug Administration (FDA) as "safe and effective", or as "new drugs" as defined by section 201(p) of the Federal Food, Drug & Cosmetic Act (the "Act") [21 U.S.C. § 321(p)], or be "legally marketed" in conformity with any over-the-counter (OTC) Drug Monograph and related considerations. All these issues are within the purview and jurisdiction of the FDA. Should any readers intend to use these botanicals, I strongly recommend that they first consult with their physician and the FDA [10903 New Hampshire Avenue, WO51 Silver Spring, Maryland 20993-0002, Website: www.fda.gov; Telephone: 1 (888) 463-6332].

Table 8.1 – Botanicals for Lyme treatment and claimed characteristics

Plant name	Compound	Indication	Activity	Other uses

Lyme disease

1. *Artemisia annua* (sweet wormwood)	Artemisinin	o Anti-parasitic o Anti-oxidant	o **Lyme bacteria** (including cyst forms) o *Babesia* (Lyme co-infection). Also: o *Schistosoma* o *Plasmodium* (malaria) o *Toxoplasma gondiio*	After 1 week, left alive ~24% Borrelia cysts. By contrast: ciprofloxacin and doxycycline left 28-49%
2. *Astragalus membranaceous* root	Phytochemicals (flavonoids, saponins)	o Immune stimulant o Anti-inflammatory o Anti-oxidant	o Breaks up **biofilm**	o Candida o Several pathogenic bacteria
3. Black walnut green huls (*Juglans nigra*)	Phytochemicals (Juglone)	o Anti-microbial o Anti-oxidant o Anti-parasitic o Anti-fungal	Kills ***Borrelia*** bacteria (spirochete, cyst, and biofilm)	Yeast *candida albicans*
4. Buckthorn bark (*Frangula alnus*)	Phytochemicals (polyphenols, flavonoids, saponins)	o Anti-oxidant o Anti-viral o Anti-bacterial o Anti-fungal	Disrupts **biofilm**	Laxative
5. Boneset (*Eupatorium perfoliatum* or fever wort or sweating plant)	Phytochemicals	o Anti-inflammatory o Anti-oxidant o Anti-bacterial o Anti-viral	Supports against seasonal viruses to help immune system in **CLD**	o Anti-cancer o Anti-malaria o Fever and cold (in Europe) o Sweat inducer
6. Cat's claw bark (*Uncaria tementosa*) Native to tropical rain forests	Phytochemicals	o Immune system helper o Anti-oxidant o Anti-viral o Anti-inflammatory	Regulates **immune function**	o Peru's "life-giving" plant o Rheumatoid arthritis (significant reduction in joints pain)
7. Cranesbill root (*Geranium maculatum*)	Phytochemicals (tannins)	o Anti-bacterial o Anti-viral o Anti-protozean parasites o Anti-oxidant	Fights ***Borrelia*** **parasites**	Geranium species for: o Cough o Diarrhea o Fever o Parasitic roundworms o Rashes
8. Devil's claw (*Harpagophytum procumbens*)	Phytochemicals	Anti-inflammatory	**Lyme's joints inflammation and pain** (especially knees): ~ 90% of cases	In Africa: o Allergies o Indigestion, and o Liver and kidney issues o Pain In other countries: o Arthritis: ~ 37% drop in knee pain
9. Essiac blend	Burdock root + Indian rhubarb root + sheep sorrel leaves + slippery elm bark	o Anti-oxidant o Prevents free radical damage to cellular DNA	Regulates/normalizes **immune responses**	o Combination of 4 herbs created by the Ojibwa tribe in Canada o Popular alternative cancer therapy

Lyme disease

				o Quenches highly reactive hydroxyl-free radicals
10. Eleuthero (root) (Siberian ginseng *senticosus*, *Acanthopanax senticosus*)	Phytochemicals	o Anti-inflammatory o Pain reducer	o **Lyme pain and inflammation** o Moves stagnant ***Borrelia*** **toxins** through lymphatic system o **Balancing effect on immune system**	o Not a true ginseng o A staple in traditional Chinese medicine (TCM) o Calms the mind o Increases energy o Strengthens spleen o Supports kidneys
11. Hawthorn berry leaf (*Crataegus monogyna* or *C. laevigata*)	Phytochemicals	o Anti-oxidant	o **Lyme carditis** (~ 10% of cases)	o Cardiovascular problems (heart failure, high blood pressure, irregular heart beat) o Reduces heart palpitations, difficulty breathing o Supports blood flow
12. Horsetail plant (*Equisetum arvense*)	Phytochemicals	o Anti-inflammatory (including reducing arthritic inflammation) o Anti-microbial	**Lyme inflammation**	o One of the oldest plant species on Earth o Strong effect on several bacteria/fungi (including *Candida albicans*)
13. Japanese knotweed root (*Fallopia Japonica* or *Polygonum cuspidatum*)	Phytochemicals (resveratrol)	o Anti-inflammatory (increased regulation of T-cells) o Anti-cancer o Anti-oxidant	Kills ***Borrelia* spirochetes**	o Considered troublesome weeds in Europe, North America, and Australia o Can boost regulatory T-cells by 47%
14. Milk thistle seed (*Silybum marianum*)	Silymarin	o Anti-inflammatory o Anti-oxidant o Anti-viral	o **Liver protection in Lyme disease** o ***Babesia***	Protects liver (more than 2000 years)
15. Nettle leaf (*Urtica dioica*)	Phytochemicals (polyphenols)	o Anti-bacterial (including *Candida albicans*) o Anti-inflammatory	o **Lyme joints pain and arthritis** o Supports **immune system**	o Known worldwide o Allergies (seasonal) o Bladder infections o Prostate enlargement o Skin rashes
16. Pau d'Arco barc (Tahebo from the *Tabebuia impetiginosa* tree)	Phytochemicals (beta-lapachione)	o Anti-inflammatory o Anti-parasitic	**Lyme inflammation**	o Native of South America's tropical rain forests o Arthritis o Fever o Fights parasites (*Leishmania*, helminths) o Pain
17. Teasel root	Phytochemicals	Anti-inflammatory	o **Lyme joints**	o Pain: back, knee,

Lyme disease

(Dipsacus asperoides)			inflammation o Directly effective against **Borrelia spirochetes** (~ 95% inhibition within 4 days)	liver, bruises o Inhibits macrophages from releasing inflammatory compounds
18. Wormwood (Artemisia absynthium)	Phytochemicals	o Anti-inflammatory o Anti-parasitic	**Parasites carrying Borrelia**	o Related to herb #1 o Like sage o As effective as the drug Praziquantel against the common intestinal tapeworm (*Hymenolepsi nana*) o Crohn's disease (80% remission within 6 weeks) o Digestive disorders o Gut inflammation
19. While willow bark (Salix alba)	Phytochemicals (Salicilin, similar to aspirin)	o Anti-fever o Anti-inflammatory	**CLD pain and inflammation**	o Known since ancient times o Back pain relief (39% complete relief)
20. Yellow dock root (Rumex crispus)	Phytochemicals (Napodin)	**Biofilm inhibitor/buster**	**Lyme detoxification** (*Borrelia* bacteria)	o Inhibits *Candida albicans* o Laxative o Liver function o Malaria combatant (*Plasmodium falciparum*)
21. Turmeric (Curcumalonga)	Phytochemicals (curcumin)	o Anti-inflammatory o Anti-oxidant o Anti-parasitic	**Lyme's joints pain and arthritis**	o Combats arthritis o 49% drop in *Schistosoma mansoni* worms

Source: Data from Drs. Todd Watts and Jay Davidson

As supplements, the above plants have unfortunately not been submitted to the rigorous testing required of pharmaceutical drugs. While many of the benefits claimed for them have been demonstrated in isolated studies, the equivalents of clinical trials are sorely needed, especially in the case of Lyme patients.

Take-home points

- Chronic Lyme disease can weaken the immune system, ignite inflammation, squelch energy, provoke pain, and trigger brain fog. It can also generate harmful free radicals, disrupt mitochondria, and overwhelm detoxification pathways.

- If antibiotics are not the best chronic Lyme disease treatment, work with your body and support the systems that have been overwhelmed. Treatment must address all the contributors to the disease in the right order: Drainage pathways, organs detoxification (colon being the last) and lymphatic system, parasites, and Lyme.

- No less than 21 different botanicals (see Table 8.1) have been identified (there may be more), each one offering unique properties and several of them acting synergistically to help: lower inflammation, reduce joints pain, fight free radical damage, break up biofilm, decrease viral load, regulate the immune system, purge parasites, support detoxification, and combat Lyme bacteria.

- As supplements, the above plants have unfortunately not been submitted to the rigorous testing required of pharmaceutical drugs. While many of the benefits claimed for them have been demonstrated in isolated studies, the equivalents of clinical trials are sorely needed, especially in the case of Lyme patients.

- Before resorting to any homeopathic treatment, I strongly recommend that you first consult with your physician and the FDA to ascertain the treatment's safety and efficacy, alleged benefits, and legal distribution.

9

Is there a vaccine against LD?

Vaccination against infection is a highly effective means to control the spread of disease in a population. In general, vaccines in common use protect against highly transmissible diseases. Their effectiveness is largely based on the generation of "herd" immunity. In the case of LD, we are dealing with a disease that is not readily transmitted from person-to-person, is vector-borne, and its risk is largely influenced by geography. Nonetheless, despite these limitations to contagion, LD has become a serious and expensive public health problem.

Before dwelling on this highly specialized subject, it would be of interest to provide a brief background.

Background

The impetus for development of a vaccine against LD gained momentum in the 1990's. Two large pharmaceutical companies had devoted considerable effort to it, leading to the approval of the first LD vaccine for human use. Double-blind, randomized, placebo-controlled clinical trials—the most rigorous type of clinical trials - were completed for each of two *Borrelia burgdorferi* vaccines manufactured by Glaxo-Smith-Kline (GSK), formerly Smith-Kline-Beecham (SKB) and Pasteur-Merieux-Connaught (PMC). Each study involved more than 10,000 volunteers from areas of the U.S. where LD is common. Both vaccines were so-called recombinant vaccines against LD that were based on a specific part of *Borrelia burgdorferi* called outer surface protein A (OspA). The vaccines were found to be 49%-68 % effective in preventing LD after two injections and 76%-92% after three injections. They were also found to be 100% effective on children. The side effects were only mild or moderate as transient adverse events. The duration of the protective immunity generated in response to the vaccines is not known. The SKB vaccine was ultimately licensed as LYMErix and approved by the FDA on 21 December 1998.

The vaccine was marketed in the U.S. between 1998 and 2002. Its entry in clinical practice was slow for a variety of reasons, including its cost, which was often not reimbursed by insurance companies. Subsequently, hundreds of vaccine recipients reported they had developed autoimmune and other side effects, which some believed were attributed to specific segments of the vaccine protein. Supported by some advocacy groups, a number of class-action lawsuits were filed against GSK, alleging the vaccine had caused these health problems. These claims were investigated by the FDA and the CDC&P, which found no connection between the vaccine and the autoimmune complaints. Further, the adverse event rate was not found to be elevated among vaccine recipients.

Despite the lack of evidence that the complaints were caused by the vaccine, sales plummeted and LYMErix was withdrawn from the U.S. market by GKS in February/April 2002. On the market for only 4 years, it was pulled in the setting of negative media coverage and fears of vaccine side effects. Several factors led to its failure. GSK then announced that even with the incidence of LD continuing to rise, sales for LYMErix declined from about 1.5 million doses in 1999 to a projected 10,000 doses in 2002. It, therefore, discontinued manufacturing the vaccine, citing insufficient consumer demand.

The fate of LYMErix was described in the medical literature as a "cautionary tale". An editorial in the fame *Nature* journal cited the withdrawal of LYMErix as an instance in which "*unfounded public fears placed pressures on vaccine developers that go beyond reasonable safety considerations*." The original developer of the OspA vaccine at the Max Planck Institute in Germany told *Nature*: "*This just shows how irrational the world can be... There was no scientific justification for the first OspA vaccine LYMErix being pulled*". This prompted the renowned vaccinologist Stanley Plotkin to publish an article in 2011 in which he called the removal of the Lyme vaccine a "public health fiasco!".

In 2018, Valneva reported positive phase I interim results for its Lyme vaccine candidate. It is also an OspA vaccine but it includes European *Borrelia* strains and lacks the region of the proteins that some had attributed to adverse events.

To summarize, as of the end of 2019, that is approximately 20 years after the withdrawal of LYMErix from the market, there is still no vaccine commercially available against LD. Further, those people who were vaccinated with it are probably no longer protected against the disease as the protection diminished over time *albeit* at an unknown rate. Enthusiasm for a subsequent product may now be founded more in basic science than in the pharmaceutical industry. Nonetheless, research is ongoing to develop new vaccines. In addition to the possible avenues to protect against LD such as interruption of transmission and infection at multiple points, current research extends well beyond simple vaccination of humans with emphasis on vaccination against the tick vector itself.

(Note: Vaccines have been formulated and approved for prevention of Lyme disease in dogs. Currently, three Lyme disease vaccines are available: (1) LymeVax, formulated by Fort Dodge Laboratories, contains intact dead spirochetes which expose the host to the organism; (2) Galaxy Lyme, Intervet-Schering-Plough's vaccine, targets proteins OspC and OspA. The OspC antibodies kill any of the bacteria that have not been killed by the OspA antibodies; and (3) Canine Recombinant Lyme, formulated by Merial, generates antibodies against the OspA protein so a tick feeding on a vaccinated dog draws-in blood full of anti-OspA antibodies, which kill the spirochetes in the tick's gut before they are transmitted to the dog.)

Vaccine development considerations

The development of protective vaccines requires the appraisal of multiple factors, both common and pathogen-specific. Given the transmission mode and antigenic variation of *Borrelia burgdorferi*, the qualities that pertain specifically to this vector-borne infection must be scrutinized. As with many pathogens, the use of whole-cell lysates (that is, the cellular debris and fluid produced by lysis) *versus* subunit antigens is a safety concern for human use.

A whole-cell lysate vaccine would induce polyclonal antibody responses to multiple antigens that would make differentiation between vaccination and infection difficult. Similarly, conserved antigens amongst spirochetes and other bacteria could confound interpretation of diagnostic tests for Lyme.

On the other hand, subunit vaccines would induce responses to a single or a few antigens, allowing easy distinction from an infection response. The issue would then be how to determine protection and efficacy, and if a serological approach would be required. For *Borrelia*, which is a pathogen with multiple species and variants found on several continents, will the vaccine protect against other genospecies or variants. Also, given the ability of the tick vector to harbor and transmit multiple pathogens concurrently upon feeding, the protection against possible co-infections must also be taken into account.

Lastly, and of significant importance, is the duration and type of immunity elicited. The generation of long-lasting B-cell memory responses to *Borrelia burgdorferi* or tick antigens would be ideal. This would limit the need for multiple booster injections to retain immunity.

Present status of the development of vaccines against LD

Several tick molecules with the potential to serve as vaccines to impair feeding and transmission have been identified in the last decade. The sequenced genome of *Ixodes scapularis* should enable the development of an effective vaccine against LD.

A tick-based vaccine holds the promise that it might be useful to also simultaneously block the transmission of other tick-borne pathogens. Technologies to genetically manipulate *Ixodes scapularis* are also coming of age, increasing our understanding of tick genes' development, feeding, and pathogen transmission. It will also help prioritize tick antigens for vaccine development.

The Connecticut Agricultural Experiment Station (CAES) and U.S. Biologic, Inc. released the publication of a field trial study showing the effectiveness of an orally-delivered anti-Lyme vaccine that targets the white-footed mouse (the major wildlife source of Lyme disease). The study took place in the residential area of Redding, CT, over a three-year time period and showed substantial decreases in the number of infected mice. One year into the study, test sites that had been treated with the vaccine showed a 13-times greater decrease in black-legged ticks *Ixodes scapularis* (the primary vector associated with the spread of disease) infected with *Borrelia burgdorferi* compared to control sites (i.e., 26% drop versus 2% drop). The vaccine causes the mice to generate antibodies and, therefore, previously infected ticks act as a "xenodiagnostic marker" of vaccine impact, meaning once they ingest the antibodies, while feeding on vaccinated mice, the ticks then become "cleared" of infection.

Reservoir-based approaches in endemic areas – case of rodents

Lyme disease

Infectious disease "reservoirs" are what scientists call the places or populations that harbor disease-causing pathogens.

Figure 9.1 - Points at which interruption of *Borrelia burgdorferi* transmission to humans can be achieved through vaccination

Direct human vaccination against *Bb* with *Bb*-specific antigens

Vector-targeted vaccines that either prevent tick feeding (tick protein) or pathogen transmission (*Bb* protein expressed in ticks or tick protein that facilitates transmission)

Reservoir-targeted vaccines (block uptake of pathogen by ticks)

Vector-targeted vaccines that either prevent tick feeding (tick protein) or pathogen colonization (*Bb* protein expressed in ticks or tick protein that facilitates colonization)

Source: Comstedt P et al (2014)

For example, certain types of wildlife can be long-term carriers, or hosts, of a disease. Rodents are a major reservoir for LD, so scientists have been looking at ways to prevent them from getting infected with *Borrelia burgdorferi*. Stopping the bacterial infection in rodents could potentially prevent transmission of the bacteria to the ticks that depend on the rodents in their early life cycle and, therefore, prevent transmission from ticks to humans. The (U.S.) National Institute on Allergies & Infectious Diseases (NIAID) is substantially contributing to these research efforts either directly by its in-house scientists or indirectly by funding other scientists.

Vaccination of the reservoir hosts and/or humans are not mutually exclusive options. In conjunction with vaccination of humans, targeting the reservoir populations to decrease tick populations and interrupting acquisition or transmission cycles should lead to the control of tick-borne pathogens. Several research groups are investigating the potential of reservoir vaccines to reduce LD in humans focusing on the three types of vaccines discussed below (Figure 9.1).

- **Oral-bait vaccines:** These vaccines are aimed at preventing mice from becoming infected, thereby interrupting the transmission cycle. Some groups are field-testing their products with support from CDC&P. Thus, an experimental vaccine-laced bait delivery system has been developed. It was tested on mice which were subsequently exposed to *Ixodes* ticks carrying multiple strains of *Borrelia burgdorferi*. Oral vaccination was found to protect 89% of the mice from infection. The blood tests showed their immune systems had created antibodies to the Lyme bacteria.

- **Rice-based vaccine elements:** In a joint study between NIAID, Ventria Bioscience, and CDC&P, rice plants that contain vaccine elements that could eventually be fed to rodent populations were grown, thus blocking the transmission cycle of the disease from rodents to ticks to people. These findings are consistent with the results reported by other investigators.

- **Use of the *Vaccinia* virus:** In other studies, in a mouse-targeted vaccine using the *Vaccinia* virus, a single oral dose resulted in strong immune system response and full protection from *Borrelia burgdorferi* infection. In addition, a significant clearance of *Borrelia burgdorferi* was observed from infected ticks who fed on vaccinated mice.

The above findings indicate that such a vaccine may effectively reduce the incidence of LD in endemic areas. The pictorial of Figure 9.1 shows points at which interruption of *Borrelia burgdorferi* (Bb) transmission to humans can be achieved through vaccination by stopping the Bb spirochetes through direct human vaccination against Bb with Bb-specific antigens. This can be achieved along three routes:

- **Vector-targeted vaccines by transmission:** They either prevent tick feeding (tick protein) or pathogen transmission (Bb protein expressed in ticks or tick protein that facilitates transmission);

- **Vector-targeted vaccines by colonization:** They either prevent tick feeding (tick protein) or pathogen colonization (Bb protein expressed in ticks or tick protein that facilitates colonization); and

Lyme disease

- **Reservoir-targeted vaccines:** They block the uptake of pathogens by ticks.

Human vaccine development

Ongoing research activities in human vaccines include the following:

- **Targeting tick saliva:** Multiple research projects are in early-stage discovery and characterization of novel vaccine formulations and targets. They include approaches that target tick saliva that is critical for the transmission of the Lyme bacteria to humans. Tick proteins that facilitate transmission of LD bacteria or that enhance survival of those bacteria in vertebrate hosts have been identified. Studies are ongoing to see if vaccines specifically targeting some of these proteins may either be used as a strategy or an "anti-tick vaccine" to be used to prevent disease. Figure 9.2 is a schematic representation of the dynamic tick saliva. *Ixodes scapularis* engorges on a vertebrate host skin for 3–7 days, spitting saliva into the host dermis at the bite-site. Salivary composition potentially changes during feeding to confront the different host defense responses.

Figure 9.2 - Temporally changing the composition of tick saliva spit into the host's skin

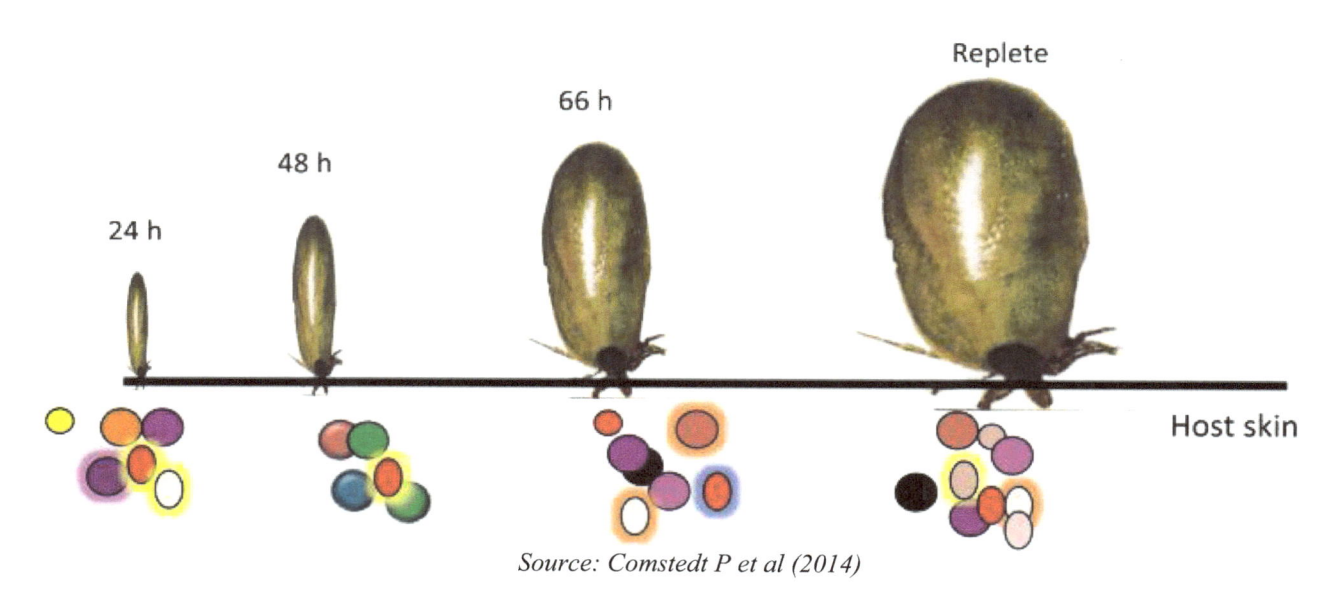

Source: Comstedt P et al (2014)

- **Modifying the successful canine Lyme disease vaccine;** and

- **Incorporating antigens against the co-infection *anaplasmosis*.**

Notwithstanding the above and other efforts, a vigorous initiative is still needed for the targeted prevention of tick-borne diseases.

The essential antibody response

Lyme disease

Studies in mice have shown that immunity by active immunization to reinfection with *Borrelia burgdorferi* is short-term and declines significantly by 1 year. Reports of human or non-human primate infection with both the LD spirochete and relapsing fever-causing spirochetes also indicate that incidental hosts are likewise susceptible to reinfection. Therefore, immune responses generated during the natural course of infection are insufficient for long-term protection.

By contrast, passive immunization with serum from acute infection in mice or chronic infection in humans has been shown to be protective. In fact, the importance of antibody responses in controlling these bacterial infections is well-established. These and other findings indicate that the *Borrelia burgdorferi* spirochetes alter antigen expression during infection so as to evade the antibody response. Further, they do not elicit effective memory responses to protective antigens (i.e., those that are expressed by all spirochetes and are likely essential for infectivity). Thus, identification of suitable antigens for induction of protective immunity has been a challenge.

For the interested reader, Sidebar 9.1 provides additional, more technical considerations on Lyme vaccine research and development.

Take-home points

- Vaccination against infection is a highly effective means to control the spread of disease in a population. In general, vaccines in common use protect against highly transmissible diseases. Their effectiveness is largely based on the generation of "herd " immunity.

- Lyme disease is a vector-borne contagion that is largely influenced by geography and is not readily transmitted from person-to-person. Nonetheless, it has become a serious and expensive public health problem.

- A vaccine was marketed in the U.S. between 1998 and 2002 but several factors led to its failure and discontinued manufacturing. This result has been considered a "public health fiasco!".

- Currently, after approximately 20 years from the withdrawal of the LYMErix vaccine from the market, there is still no vaccine against LD and those people who were vaccinated are probably no longer protected as the protection diminishes over time.

- There are several possible avenues regarding new vaccine development such as interruption of transmission and infection at multiple points, vaccination against the tick vector itself, etc.

- In the development of protective vaccines, the qualities that pertain specifically to *Borrelia burgdorferi* vector-borne infection must be scrutinized. A whole-cell lysate vaccine would make differentiation between vaccination and infection difficult. Similarly, conserved antigens amongst spirochetes and other bacteria could confound interpretation of diagnostic tests for Lyme. On the other hand, subunit vaccines would induce responses to single or few antigens, allowing easy distinction from an infection response. The issue remains as to how to determine protection and efficacy, and if a serological approach would be required.

Lyme disease

- The duration and type of immunity elicited are significantly important. The generation of long-lasting B-cell memory responses to *Borrelia burgdorferi* or tick antigens would be ideal and would limit the need for multiple booster injections to retain immunity.

- Several tick molecules with the potential to serve as vaccines to impair feeding and transmission have been identified. Further, the sequenced genome of *Ixodes scapularis* should enable the development of an effective vaccine against LD.

- A tick-based vaccine holds the promise that it might be useful to also simultaneously block the transmission of other tick-borne pathogens.

- Infectious disease reservoirs are places or populations that harbor disease-causing pathogens. Stopping the bacterial infection in rodents could potentially prevent transmission of the bacteria to the ticks that depend on the rodents in their early life cycle and prevent transmission from ticks to humans.

- Research and development of reservoir vaccines in mice focus on oral bait vaccines (for interrupting the transmission cycle), rice-based vaccine elements (for blocking the transmission from rodents to ticks to people), and use of the *Vaccinia* virus (for eliciting a strong immune system response and full protection from *Borrelia burgdorferi* infection).

- Ongoing research activities in human vaccine development include targeting tick saliva (critical for transmission to humans), modifying the successful canine Lyme disease vaccine for use in humans, and incorporating antigens against the co-infection *anaplasmosis*.

- Notwithstanding the above and other efforts, a vigorous initiative is needed for the targeted prevention of tick-borne diseases.

- Studies in mice have shown that immunity to reinfection with *Borrelia burgdorferi* is short-term and declines significantly by 1 year. Immune responses generated during the natural course of infection are insufficient for long-term protection. By contrast, passive immunization with serum from acute infection in mice or chronic infection in humans is protective.

- The *Borrelia burgdorferi* spirochetes alter the antigen expression during infection to evade the antibody response and do not elicit effective memory responses to protective antigens, leaving the identification of suitable antigens for induction of protective immunity as a remaining challenge.

- Given the ability of the tick vector to harbor and transmit multiple pathogens concurrently upon feeding, protection against possible co-infections must be taken into account as well as the duration and type of immunity elicited.

- Several tick molecules with the potential to serve as vaccines have been identified to impair feeding and transmission. Genome sequencing of the *Ixodes scapularis* will enable the

Lyme disease

development of an effective vaccine against Lyme disease.

- Vaccination of the reservoir hosts and/or humans are not mutually exclusive options, and targeting the reservoir populations to decrease tick populations and interrupting acquisition or transmission cycles in conjunction with vaccination of humans should provide the desired goal of controlling tick-borne pathogens.

Sidebar 9.1
Further considerations on Lyme vaccine research and development

The three main strategies for developing a vaccine are: (a) the transmission-blocking vaccine, (b) targeting the reservoir host, and (c) targeting the tick vector. Each of these is discussed below.

Strategy # 1 - On the transmission-blocking vaccine and its demise

Several antigen subunits of *Borrelia burgdorferi* have been evaluated for their vaccine potential (see Table S9.1). Except for OspA, all antigens listed are not vaccine candidates on their own. OspA is a lipoprotein whose expression is abundant on *in vitro*-cultured spirochetes and spirochetes within the tick midgut. It is also quite immunogenic and immunization with OspA provides cross-protection of mice challenged with the North American isolates of *Borrelia burgdorferi*.

Table S9.1 - Prospective Lyme vaccine antigens from *Borrelia burgdorferi*

Borrelia burgorferi antigen	Protective mechanism	How tested?	Result	References
OspA	Antibody-mediated transmission blocking	o Challenge of mice by infection, tissue transplant, and transmission o Challenge of monkeys by tick transmission	Efficacious. Dependent upon antibody titer	Fikrig *et al.* (1990, 1992) Philipp *et al* (1997) Probert and Lefebvre (1994) Telford *et al* (1995)
OspB	o Antibody mediated o Elicits bactericidal antibodies	Active and passive protective against infection challenges	Potential for strain-dependent efficacy due to truncations of OspB proteins in some strains	Coleman *et al* (1994) Fikrig *et al.* (1993) Probert and Lefebvre (1994) Probert *et al* (1997) Telford *et al* (1993)
OspC	Antibody-mediated within host	Challenge of mice by injection and tick transmission	o Effective but with minimal cross-species protection o Failure to elicit long-term (anamnestic) response	Gilmore *et al* (1996, 2003) Probert and Lefebvre (1994) Probert *et al* (1997)

DbpA	Antibody-mediated within host	Challenge of mice by injection and tick transmission	Protective against injected, but not tick-transmitted infection	Hagman *et al* (2000) Hanson *et al* (1998)
Bbk32 (p35)	Antibody-mediated within host	Passive immunization against infection and tick challenge	Efficacy in combination with DbpA and OspC against challenge by infection and not singly	Brown *et al* (2005) Fikrig *et al.* (1997, 2000)

Source: Adapted from Comstedt P et al (2014)

Some of the positive and negative aspects of the OspA vaccine can be found in Table S9.2. However, reports emerged suggesting that the vaccine could induce arthritis. This led to anti-vaccine sentiment and class action lawsuits, along with reduced support amongst physicians for the vaccine and eventually enough of a decline in use for its voluntary removal by the manufacturer. Unfortunately, this failure in North America led the leading European prospective Lyme vaccine manufacturer, Pasteur-Merieux-Connaught, to halt its development. This further led to the demise of the OspA vaccine.

Table S9.2 - Positive and negative characteristics of the OspA vaccine

Positive	Negative
o Blocks transmission o Easier to test for efficacy	Requires maintenance of high antibody titers for efficacy (multiple boosts)
Subunit does not interfere with immunodiagnosis	o Some adverse reactions o Potential for reduced autoimmunity
Targets a reasonably-conserved protein within species	Not effective against other tick-borne diseases

Strategy # 2 - Targeting the reservoir (mouse) host

Another option for interrupting the transmission of *Borrelia burgdorferi* to humans is through the vaccination of reservoir hosts. The majority of human infections are transmitted by *Ixodes* ticks in the nymphal stage so blocking acquisition at the pre-nymphal (larval) stage would be most effective for preventing human infection.

Considerations for a reservoir host vaccine include the antigen type, the route of delivery, the type of delivery system, and the implementation protocol. The OspA antigen is the most efficacious vaccine in animals and the primary choice for the first reservoir-targeting vaccine strategies. It acts by blocking transmission, Vaccination has an impact on the percentage of infected ticks the following year so targeting mouse-dense areas can have a significant impact on carriage, however, the contribution of non-mouse species must also be considered.

The other approach, the baited oral vaccination strategy, achieved protection of mice (89%) and reduction of *Borrelia burgdorferi* in vector ticks. The advantages of such an approach are its efficacy and the absence of safety issues.

In a third approach, OspA was delivered by the *Vaccinia* virus (VV) for several reasons: (1) these viruses have a broad host range; (2) they are stable under the harsh conditions encountered in the digestive tract; (3) they can express proteins at high levels from only a single dose; and (4) their ingestion does not cause disease in wildlife nor is it readily transmissible amongst infected animals. However, the potential to transmit the virus to unwanted recipients remains.

Strategy # 3 - Targeting the tick vector

Historically, vaccines against infectious agents, including *Borrelia burgdorferi*, have primarily utilized live attenuated pathogens or antigens of the pathogen to induce protective immunity. A potent alternative avenue to protect against arthropod-borne pathogens is targeting the vector itself, be it to eliminate the vector by using chemicals toxic to that vector, by para-transgenic approaches that modify the vectors' ability to transmit pathogens or reproduce, or by use of vaccines targeting vector antigens critical for the vector to feed, reproduce or transmit pathogens. *Borrelia burgdorferi* is transmitted by five species of *Ixodes* ticks within the *Ixodes ricinus* complex: *Ixodes scapularis, Ixodes pacificus,* and *Ixodes cookei* in North America, and *Ixodes ricinus* and *Ixodes persulcatus* in Europe and Asia, respectively. Additionally, *Ixodes scapularis* transmits *Anaplasma phagocytophilum, Babesia microti,* and Powassan virus in North America while *Ixodes ricinus* and *Ixodes persulcatus* transmit tick-borne encephalitis virus in Europe and Asia.

Acquired resistance to ticks – just a matter of time

Several decades ago, it was observed that rabbits infested repeatedly with *Dermacentor* ticks developed a robust immune response resulting in rapid rejection of ticks. This phenomenon of acquired tick resistance has also been noted in various tick-host models. *Ixodes scapularis* ticks feed successfully on guinea pigs and rabbits at first infestation, but feeding is reduced and ticks fall-off or die within 12–24 hours at subsequent infestations. The hallmark of tick resistance is the swelling and redness at the tick bite site due to cutaneous basophil hypersensitivity, or the rapid recruitment of basophils to the tick bite-site which, followed by their degranulation, effectively thwarts tick feeding, and promotes tick mortality. It is presumed that salivary proteins secreted into the bite site provoke the immune response in the host that recruits basophils to the site. Hence, there is an ongoing interest to exploit the phenomenon of acquired tick resistance to identify tick salivary proteins that are natural targets of host immunity. This would help define salivary protein candidates that might serve as vaccine targets to block tick feeding and *Borrelia* transmission.

Blocking *Borrelia* transmission – the real deal

Blocking tick feeding might just be a monumental task, up against the powerful evolutionary measures designed to ensure that the tick saliva is equipped with protein and non-protein biomolecules critical for feeding. But, from a human vaccine perspective, do we really want to block tick feeding? Is it not sufficient that we block pathogen transmission? A vaccine that can effectively block pathogen transmission is undoubtedly the public health goal that is broadly applicable to the murine reservoir host and to humans. Whereas the manufacture of the first such vaccine (OspA) has been discontinued, a safe and effective vaccine against LD has since then remained an unmet need.

10

Prevention and prognosis

Prevention

Tick bites may be prevented by avoiding or reducing time in likely tick habitats and taking precautions while in and when getting out of one such habitat.

Most Lyme human infections are caused by *Ixodes* nymph bites between April and September. Ticks prefer moist, shaded locations in woodlands, shrubs, tall grasses, and leaf litter or wood piles. Tick densities tend to be highest in woodlands, followed by unmaintained edges between woods and lawns (about half as high), ornamental plants and perennial ground cover (about a quarter), and lawns (about 30 times less).

Ixodes larvae and nymphs tend to be abundant also where mice nest, such as stone walls and wood logs. They typically wait for potential hosts ("quest") on leaves or grasses close to the ground with forelegs outstretched. When a host brushes against its limbs, the tick rapidly clings and climbs on the host looking for a skin location to bite.

In Northeastern United States, 69% of tick bites are estimated to happen in residences, 11% in schools or camps, 9% in parks or recreational areas, 4% at work and in hunting areas, and 3% while hunting. Activities associated with tick bites around residences include yard work, brush clearing, gardening, playing in the yard, and letting into the house dogs or cats that roam outside in woody or grassy areas. In parks, tick bites often happen while hiking or camping. Walking on a mowed lawn or center of a trail without touching adjacent vegetation is less risky than crawling or sitting on a log or stone wall. Pets should not be allowed to roam freely in likely tick habitats.

Precautionary measures that kill ticks

As a precaution, CDC&P recommends soaking or spraying clothes, shoes, and camping gear such as tents, backpacks, and sleeping bags with 0.5% *permethrin* solution and hanging them to dry before use. Permethrin is odorless and safe for humans but highly toxic to ticks. After crawling on permethrin-treated fabric for as few as 10–20 seconds, tick nymphs become irritated and fall off or die. Permethrin-treated closed-toed shoes and socks reduce by 74 times the number of bites from nymphs that make first contact with a shoe of a person also wearing treated shorts (because nymphs usually quest near the ground, this is a typical contact scenario). Better protection can be achieved by tucking permethrin-treated trousers (pants) into treated socks and a treated long-sleeve shirt into the trousers so as to minimize gaps through which a tick might reach the wearer's skin. Light-colored clothing may make it easier to see ticks and remove them before they bite. Military and outdoor workers' uniforms treated with permethrin have been found to reduce the number of bite cases by 80%-95%. Permethrin protection lasts several weeks of wear and washings in customer-treated items and up to 70 washings for factory-treated items. It should not be used on human skin, underwear or cats (Table 10.1).

Precautionary measures that repel ticks

Unlike permethrin, repellents repel but do not kill ticks, protect for only several hours after application, and may be washed off by sweat or water. The (U.S.) Environmental Protection Agency (EPA) recommends several tick repellents for use on exposed skin, including DEET, picardin, IR3535 (a derivative of the amino acid beta-alanine), oil of lemon eucalyptus (OLE, a natural compound) and OLE's active ingredient para-menthane-diol (PMD). In the U.S., the most popular repellent is DEET whereas in Europe it is picaridin. Unlike DEET, picaridin is odorless and less likely to irritate the skin or harm fabric or plastics. Repellents with higher concentration may last longer but are not more effective. Against ticks, 20% picaridin may work for 8 hours vs. 11% DEET for 5–6 hours or 30%-40% OLE for 6 hours. Repellents should not be used under clothes, on eyes, mouth, wounds or cuts, or on babies younger than 2 months (3 years for OLE or PMD). If sunscreen is used, repellent should be applied on top of it. Repellents should not be sprayed directly on a face, but should instead be sprayed on a hand and then rubbed on the face (Table 10.1).

Table 10.1 – Tick killers and repellents

Category	Product	Concentration – Acting time	Notes
Killer	Permethrin	5%	
Repellent	o DEET o OLE o Picaridin o IR3535	o 11% - 5 to 6 hours o 30%-40% - 6 hours o 20% - 8 hours	o Preferred in the U.S. o Preferred in Europe

After coming indoors, clothes, gear, and pets should be checked for ticks. Clothes can be put into a hot dryer for 10 minutes to kill ticks (just washing or warm-drying are not enough). Showering as soon as possible, looking for ticks over the entire body, and removing them reduce risk of infection. Unfed tick nymphs are the size of a poppy seed, but a day or two after biting and attaching themselves to a person, they look like a small blood blister. The following areas should be checked especially carefully: armpits, between legs, back of knee, bellybutton, trunk, and in children's ears, neck, and hair.

Sidebar 10.1 summarizes the tick-bite prevention you can take for you, your family, and your pets. Sidebar 10.2 illustrates how to remove attached ticks if you have already been bitten.

Prognosis

In the U.S., several ticks carry pathogens that can cause human diseases including: *anaplasmosis; babesiosis; Borrelia miyamotoi*: Bourbon virus: Colorado tick fever: *ehrlichiosis:* heartland virus: Lyme disease: Powassan disease: *Rickettsia parkeri rickettsiosis;* Rocky Mountain spotted fever; Southern tick-associated rash illness (STARI): tick borne relapsing fever; *tularemia:* and 364D *rickettsiosis* (or *Rickettsia phillipi*).

As I have extensively discussed in the previous Chapters, while *Borrelia* is responsible for causing primary LD, it is usually not the only factor. When one or several ticks perhaps from different species do bite, various bacteria and parasites also enter the wound causing other opportunistic infections. Many of these other infections complicate treatment and increase the risk for autoimmune diseases depending on the patient's genetic background. They can also be a precursor to various cancers and even neurodegenerative diseases. Treating these infections in their totality as quickly as possible is essential to full recovery and to achieve health. Chronic Lyme disease (CLD) or syndrome is the overall disease caused by both the primary and secondary infections. Its signs of illness may appear gradually over time or may have never entirely subsided with earlier treatment. Further, the infection can impair or even suppress the immune system so the body does not create enough antibodies to fight the infecting bacteria. CLD can cause a significant burden on patients more so than Lyme disease alone. It can linger for months or even years, typically worsening over time in the absence of proper treatment and leading to debilitating symptoms. Post-treatment Lyme disease syndrome (PLDS) describes the continued ongoing resistance to antibiotics and continued symptoms.

Lyme patients who are diagnosed early, and receive proper antibiotic treatment, usually recover rapidly and completely. However, even after successful treatment with antibiotics and resolution of their clinical symptoms, some patients can still have the bacteria in their blood for decades. Although most cases of LD can be cured with a 2- to 4-week course of oral antibiotics, patients can sometimes have symptoms of pain, fatigue, or difficulty thinking that last for more than 6 months after they finish treatment. This may be caused by one or a combination of autoimmunity, other persistent non-Lyme infection, or even other unrelated causes. After several months, untreated or inadequately treated people may go on to develop late-disseminated infection as chronic symptoms that affect many parts of the body (brain, eyes, heart, joints, and nerves). Arthritis occurs in up to 60% of untreated people (hip, joints (large, temporomandibular), knee). Without treatment, swelling and pain typically resolve over time but periodically return. Baker cysts may also form and rupture.

Chronic neurologic symptoms occur in up to 5% of untreated people (peripheral neuropathy or polyneuropathy, encephalopathy, depression, and fibromyalgia). Chronic progressive encephalomyelitis resembling multiple sclerosis may also develop (cognitive impairment, brain fog, migraines, balance issues, weakness in the legs, awkward gait, facial palsy, bladder problems, vertigo, and back pain). Psychosis may also develop in rare cases (panic attacks, anxiety, delusional behavior including somatoform delusions, and *acrodermatitis chronica atrophicans* (a chronic skin disorder).

Lyme disease patients are often caught in a vicious cycle of immune depression that begins with the initial infection. The co-infections release a multitude of endotoxins, neurotoxins, and biotoxins that confuse the immune system and cause it to potentially attack its own cells resulting in autoimmune-like symptoms and chronic inflammation throughout the brain and body. Further, the confusion caused by toxins, and other factors depress the immune system. With the immune system severely impaired by Lyme and its co-infections, the traditional diagnostic tests may not be accurate. Worse, co-infections, biotoxins, endotoxins, mycotoxins, neurotoxins, autoimmune attacks, and opportunistic secondary infections can produce a multitude of symptoms making it difficult to recognize Lyme. It is the combination of these multiple (primary and secondary) infections with other complications that contribute to the patients' debilitating set of physical and neurological symptoms. Correctly identifying and vigorously treating all these infections is the key in producing lasting results. There also are various confounding diseases that need to be differentiated against, including spider webs, cellulitis, shingles, facial palsy, Lyme lymphocytic meningitis, Lyme radiculopathy. diverticulitis, and coronary artery syndrome. Late-stage Lyme disease may be misdiagnosed as autoimmune and neurodegenerative diseases, Crohn's disease, chronic fatigue syndrome, fibromyalgia, HIV, lupus, multiple sclerosis, and rheumatoid arthritis.

Take-home points

- Tick bites may be prevented by avoiding or reducing time spent in likely tick habitats and taking precautions while in and when getting out of one such habitat.

- In Northeastern United States, 69% of tick bites are estimated to happen in residences, 11% in schools or camps, 9% in parks or recreational areas, 4% at work and in hunting areas, and 3% while hunting.

- Several precautionary measures can be taken to repel or/and kill ticks.

- In the U.S., several ticks carry pathogens that can cause human diseases. When one or several ticks perhaps from different species do bite, various bacteria and parasites also enter the wound causing other opportunistic infections, complicating treatment and increasing the risk of other diseases.

- Lyme patients who are diagnosed early, and receive proper antibiotic treatment, usually recover rapidly and completely. However, some patients can still have the bacteria in their blood for decades.

- Chronic neurologic symptoms occur in up to 5% of untreated people. Chronic progressive encephalomyelitis resembling multiple sclerosis may also develop. Psychosis may also develop in rare cases.

- Lyme disease patients are often caught in a vicious cycle of immune depression that begins with the initial infection.

- The co-infections release a multitude of biotoxins, endotoxins, mycotoxins, and neurotoxins, that confuse the immune system and cause autoimmune diseases as well as chronic inflammation throughout the brain and body.

Lyme disease

It is the combination of these multiple (primary and secondary) infections with other complications that contribute to the patients' debilitating set of physical and neurological symptoms.

Sidebar 10.1
Measures for preventing tick bites

Tick bites must be prevented on you and your family, and your pets. In examining each of these instances, I have borrowed largely from CDC&P publications as a public service for the greater dissemination of the information they include.

Preventing tick bites on people

Tick exposure can occur year-round, but ticks are most active during certain months of the year (the warmer months of April through September), and depending on your geographical locations. (See also Chapters 17 and 18.)

So, before venturing outside and spending time outdoors, including your own backyard:

- *Know where to expect and avoid contact with ticks* (grassy, brushy, or wooded areas, or even on animals).

- *Treat or buy pre-treated clothing and gear:* With 0.5% permethrin.

- **Use Environmental Protection Agency (EPA)-registered insect repellents** and follow *product instructions:* Repellents may contain DEET, picaridin, IR3535, Oil of Lemon Eucalyptus (OLE), para-menthane-diol (PMD), or 2-undecanone. (Products containing OLE or PMD should not be used on children under 3 years.)

After returning indoors:

- *Check clothing for ticks.*

- *Remove and kill ticks in a washer/dryer:* Use hot water and 10 or more minutes tumble dry.

- *Examine gear:* Boots, coats, daypacks, etc.

- *Examine pets.*

- *Shower soon (within 2 hours) of returning indoors:* Wash-off unattached ticks.

- *Do a full body exam for attached ticks:* See Figure S10.1 for areas to be inspected.

Book Depository
bookdepository.com

These artworks were inspired by

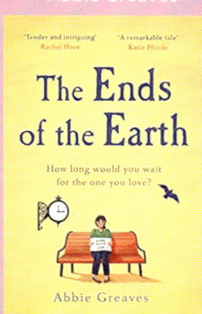

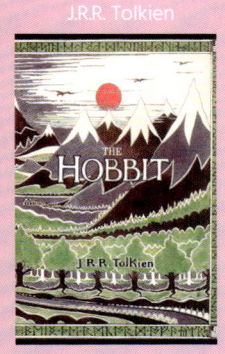

The Ends of the Earth
Abbie Greaves

ISBN13: 9781529123968

The Hobbit
J.R.R. Tolkien

ISBN13: 9780261103283

Follow us on social media

FREE DELIVERY ON 20 MILLION BOOKS
bookdepository.com

Book Depository

bookdepository.com

Image: @ashiejayn

Image: @blondiestudyblr

Our #READCreate
Competition Winners

See overleaf

Figure S10.1 – Body parts inspection for attached ticks
Source: CDC&P

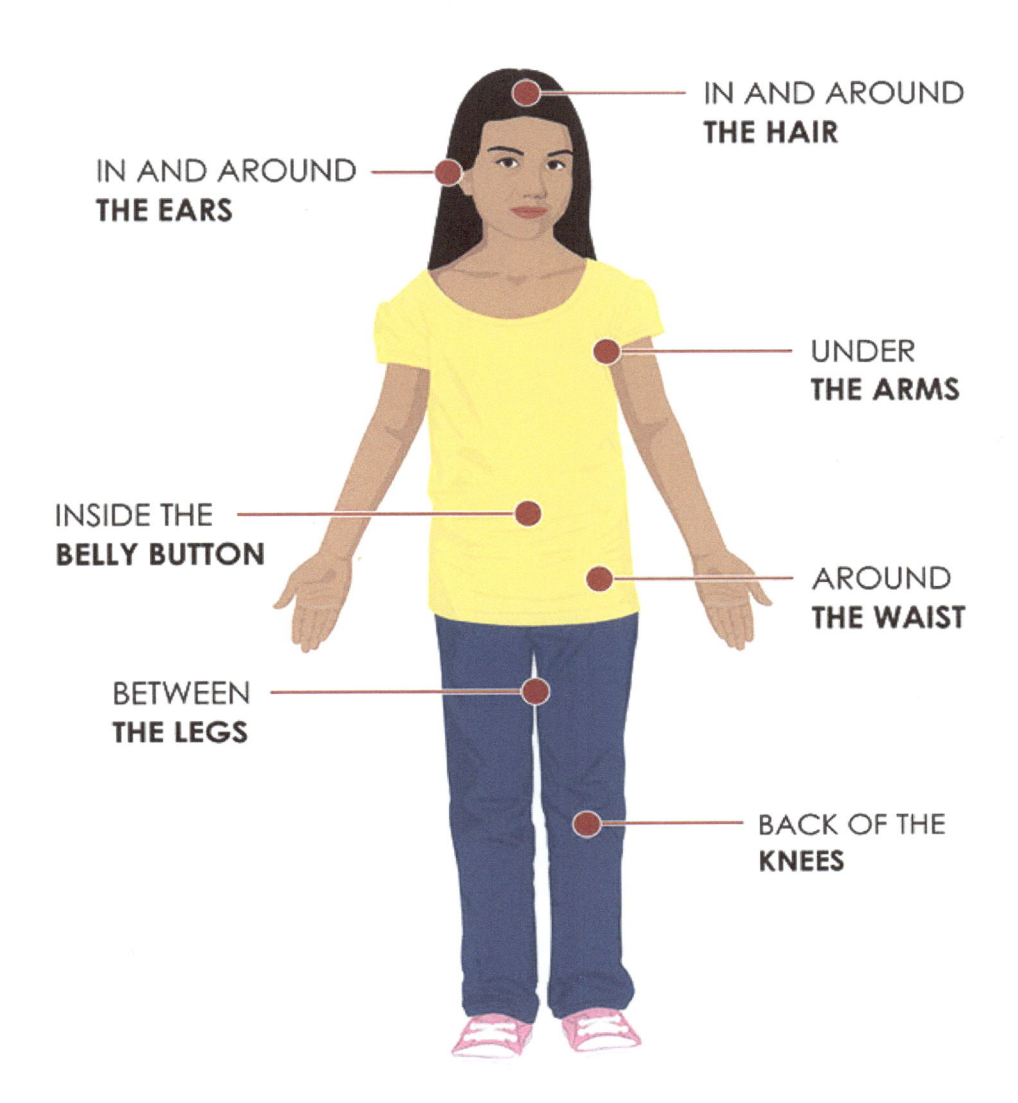

- IN AND AROUND THE HAIR
- IN AND AROUND THE EARS
- UNDER THE ARMS
- INSIDE THE BELLY BUTTON
- AROUND THE WAIST
- BETWEEN THE LEGS
- BACK OF THE KNEES

Preventing tick bites on pets

On September 2018, the FDA issued its document titled "FDA Fact Sheet for Pet Owners and Veterinarians" about potential adverse events associated with "*Isoxazoline* Flea and Tick Products".

Case of dogs: Dogs are very susceptible to tick bites and tick-borne diseases, and can bring them into the home. Signs of tick-borne disease may not appear for 7-21 days or longer after a tick bite. Furthermore, vaccines are not available for most of the tick-borne diseases that dogs can get, so:

Lyme disease

Talk to your veterinarian about: Tick-borne diseases in your area, best preventive products.

Use a tick-preventive product on your dog.

Watch for changes in behavior or appetite.

Do a full body exam for attached ticks: See Figure S10.2 for areas to be inspected.

Figure S10.2 – Dog body parts inspection for attached ticks

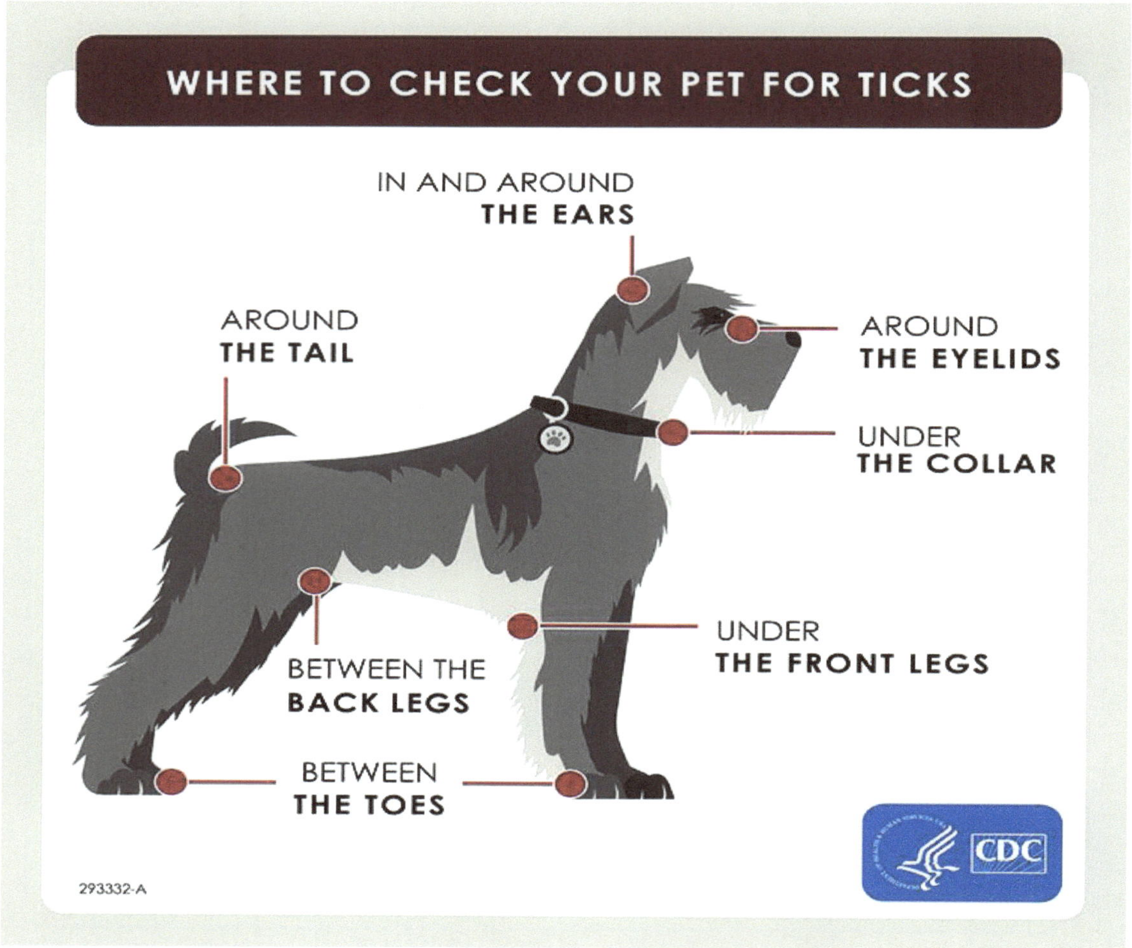

WHERE TO CHECK YOUR PET FOR TICKS

- IN AND AROUND THE EARS
- AROUND THE TAIL
- AROUND THE EYELIDS
- UNDER THE COLLAR
- UNDER THE FRONT LEGS
- BETWEEN THE BACK LEGS
- BETWEEN THE TOES

Case of cats: Cats are extremely sensitive to a variety of chemicals. Do not apply any tick prevention products to them without first asking your veterinarian.

Create a tick-safe zone in your yard with landscaping

Lyme disease

The Connecticut Agricultural Experiment Station has issued its "Tick Management Handbook" for a comprehensive guide to preventing ticks and their bites through landscaping. The following landscaping techniques can help reduce tick populations in your yard (Figure S10.3).

Figure S10.3 – Using landscaping to create a tick-safe zone in your yard

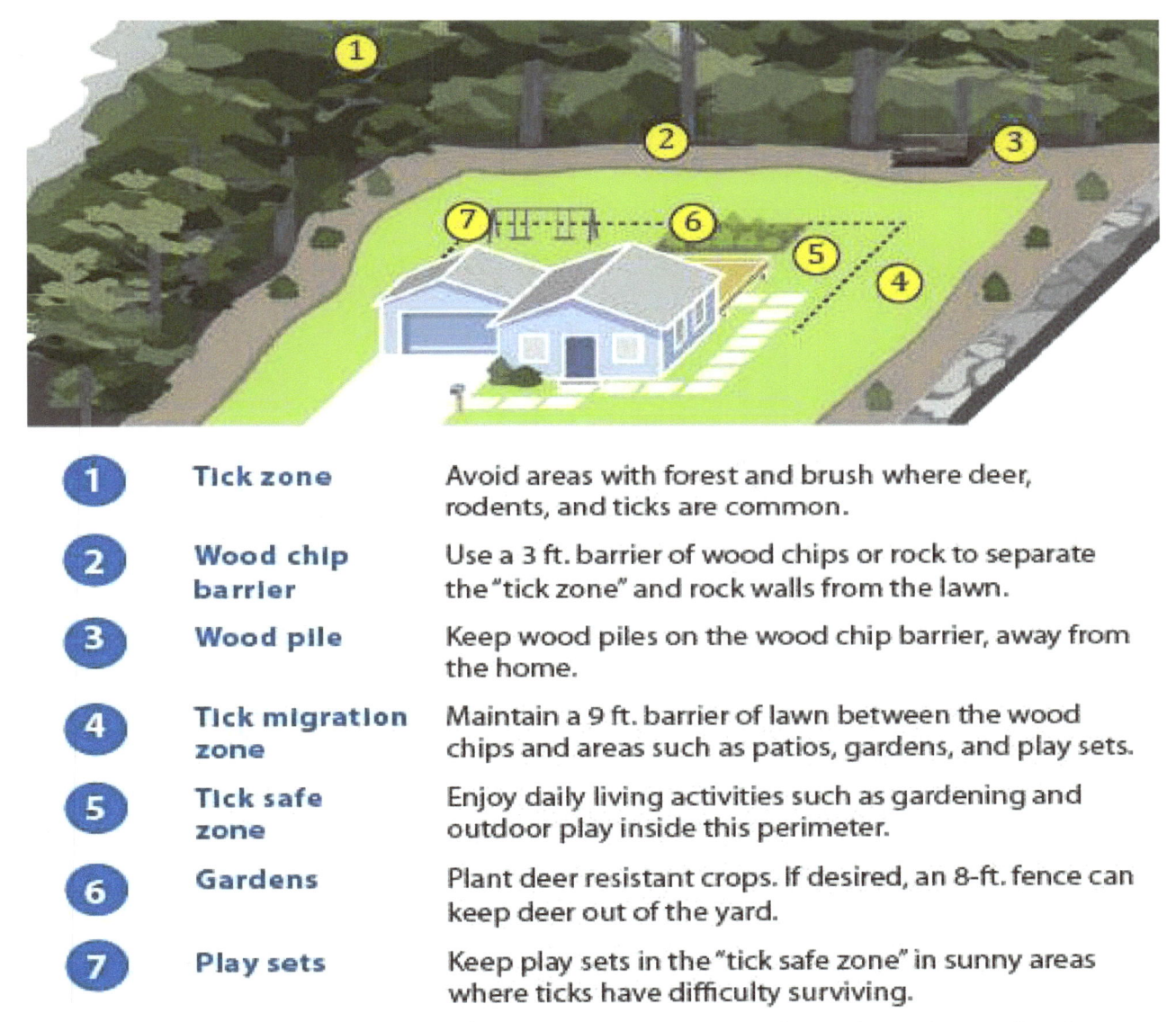

1	Tick zone	Avoid areas with forest and brush where deer, rodents, and ticks are common.
2	Wood chip barrier	Use a 3 ft. barrier of wood chips or rock to separate the "tick zone" and rock walls from the lawn.
3	Wood pile	Keep wood piles on the wood chip barrier, away from the home.
4	Tick migration zone	Maintain a 9 ft. barrier of lawn between the wood chips and areas such as patios, gardens, and play sets.
5	Tick safe zone	Enjoy daily living activities such as gardening and outdoor play inside this perimeter.
6	Gardens	Plant deer resistant crops. If desired, an 8-ft. fence can keep deer out of the yard.
7	Play sets	Keep play sets in the "tick safe zone" in sunny areas where ticks have difficulty surviving.

Based on a diagram by K. Stafford, Connecticut Agricultural Experiment Station

- Clear tall grasses and brush: around the home and at the edge of lawns.

- Place a 3-ft wide barrier of wood chips or gravel between lawns and wooded areas and around patios and play equipment: This will restrict tick migration into recreational areas.

Lyme disease

- Mow the lawn frequently and keep leaves raked.

- Stack wood neatly and in a dry area: This will discourage rodents that ticks feed on.

- Keep playground equipment, decks, and patios away from yard edges and trees: Place them in a sunny location, if possible.

- Remove all tick-hiding places: Any old furniture, mattresses, or trash.

- Apply acaricides (tick pesticides) outdoors to control ticks: After checking with local health officials about the best time to apply and identify rules and regulations related thereto.

- Consider employing a professional pesticide company: To apply pesticides at your home.

Sidebar 10.2
Removing attached ticks

If you find a tick attached to your skin, follow the procedure below to reduce your chances of getting sick and how to get treatment promptly if you do get sick:

Initial procedure

- Remove the tick as quickly as possible with fine-tipped tweezers and do not wait for it to detach;

- Grasp the tick as close to the skin's surface as possible;

- Pull upward with steady, even pressure (twisting or jerking the tick may cause the mouth-parts to break off and remain in the skin; remove them or, if unable, leave them alone);

- Thoroughly clean the bite area and your hands with rubbing alcohol or soap and water; and

- If you would like to bring the live tick to your health care provider, dispose of it by putting it in alcohol, placing it in a sealed bag/container, and wrapping it tightly in tape. Otherwise, flush it down the toilet. Never crush a tick with your fingers.

Follow-up

- If you develop a rash or fever within several weeks of removing a tick, see your doctor.

Lyme disease

- Avoid folklore remedies such as "painting" the tick with nail polish or petroleum jelly, or using heat to make the tick detach from the skin.

Figure S10.4 – Procedure for removing attached ticks

Consider calling your health care provider

Watch for symptoms for 30 days and call your healthcare provider if you get any of the following:

- Fatigue;
- Fever;
- Headache;
- Joint swelling and pain.
- Muscle pain; and
- Rash.

Treatment for tick-borne diseases

Treatment should be based on symptoms, history of exposure to ticks, and in some cases, blood test results. Most tick-borne diseases can be treated with a short course of antibiotics. In general, CDC&P does not recommend taking antibiotics after tick bites to *prevent* tick-borne diseases. However, in certain circumstances, a single dose of the antibiotic doxycycline after a tick bite may lower your risk of Lyme disease.

Common questions after a tick bite

Lyme disease

Should I get my tick tested for germs?

Some commercial companies offer to test ticks for specific germs. The results of such tests may not be reliable because laboratories that test ticks are not required to meet the same quality standards as laboratories used by clinics or hospitals for patient care. In fact, CDC&P strongly discourages using such tests when deciding whether to use antibiotics after a tick bite. Thus:

> *<u>Positive results:</u>* Can be misleading for, even if a tick contains a germ, it does not mean that you have been infected by that germ. Also:

> *<u>Negative results:</u>* Can also be misleading for you might have been bitten unknowingly by a different infected tick.

Can I get sick from a tick that is crawling on me but has not yet attached?

If not attached or full of blood, it could not have spread its germs. For this, it must bite you, feed on your blood, and then spread its germs. However, if a tick is crawling on you, there may be others and a careful tick check of your body (see Figure S10.1) is recommended.

How long does a tick need to be attached before it can spread infection?

Depending on the type of tick and germ, a tick needs to be attached to you for different amounts of time (minutes to days) to spread infection. The risk for Lyme disease is very low if the tick has been attached for fewer than 36 hours. It is recommended to check for ticks daily and remove them as soon as possible.

11

Is STARI the same as Lyme?

There is more to say about rashes without them being LD. For example:

A small bump or redness at the site of a tick bite that occurs immediately and resembles a mosquito bite is common. This irritation generally goes away in 1-2 days and is not a sign of LD.

A rash with a very similar appearance to Lyme's *erythema migrans* (EM) occurs with Southern Tick-Associated Rash Illness (STARI), but is not LD. Further, STARI rashes may take many forms.

Many other rashes are associated with the ticks listed in Table 2.1 (which include STARI).

but, then, ... what is STARI?

STARI is a rash similar to the LD rash. It has been described in humans following bites of the lone star tick, *Amblyomma americanum* (see Table 2.2). The adult female is distinguished by a white dot or "lone star" on her back. All three life stages (lymph, larva, adult) aggressively bite people.

Lone star tick may be accompanied by fatigue, fever, headache, and muscle and joint pains. It is a concern, but not for LD. Many people, even health care providers, can be confused about whether the lone star tick causes LD. It does not! Patients bitten by lone star ticks will occasionally develop a circular rash similar to the rash of early LD. The cause of this rash has not been determined but it has been shown not to be caused by *Borrelia burgdorferi*, the bacterium that causes LD. In the cases of STARI studied to date, the rash and accompanying symptoms have resolved following treatment with an oral antibiotic (doxycycline), but it is unknown whether this medication only sped recovery. Lone star ticks do not transmit *Borrelia burgdorferi*. In fact, their saliva has been shown to kill *Borrelia*.

In the U.S., the lone star tick is found throughout the eastern, southeastern, and south-central States. It has been recorded in large numbers as far north as Maine and as far west as central Texas and Oklahoma eastward, and along the Atlantic coast as far north as Maine. Its distribution, range, and abundance have increased over the past 20-30 years (Figure 11.1). All three life stages (larva, nymph, adult) of the lone star tick will feed on humans, and may be quite aggressive. They will also feed

Lyme disease

readily on animals including dogs and cats, and may be brought into the home on pets.

Figure 11.1 - Estimated geographic distribution of lone star tick

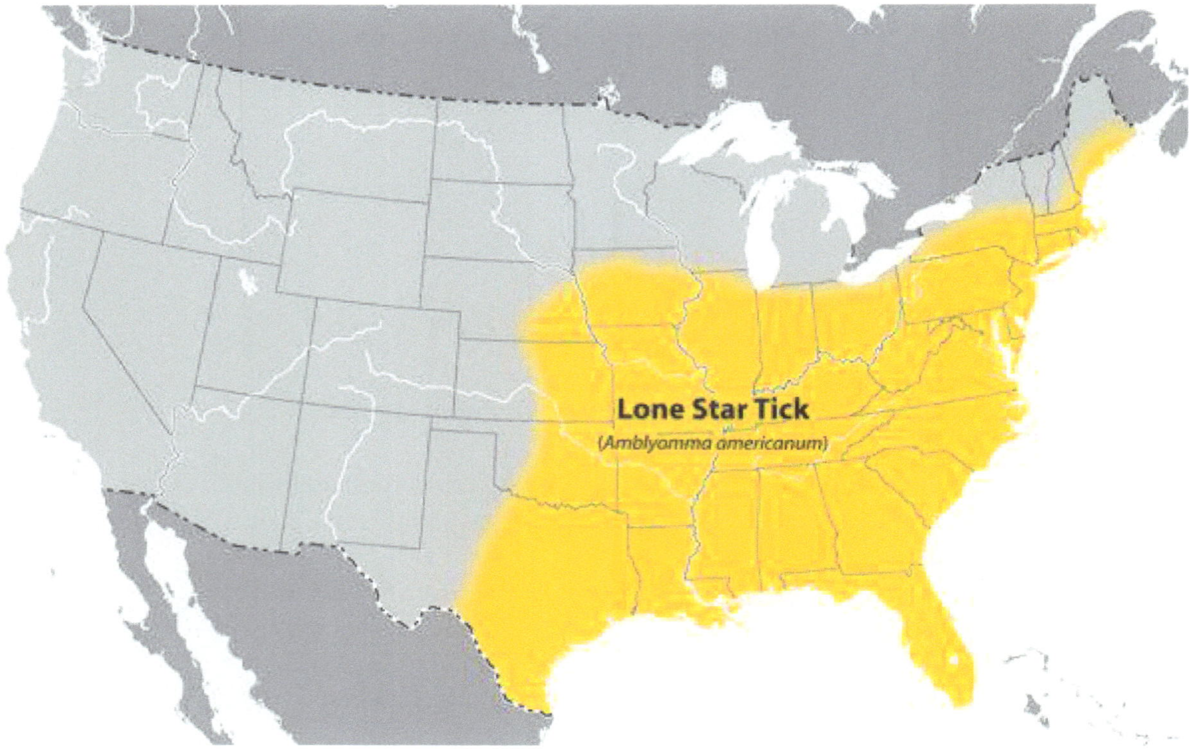

Source: CDC&P

Symptoms

The rash of STARI is a red, expanding "bull's-eye" lesion that develops around the site of a lone star tick bite. It usually appears within 7 days of a tick bite and expands to a diameter of 8 centimeters (3 inches) or more. It should not be confused with much smaller areas of redness and discomfort that can occur commonly at the site of any tick bite. Patients may also experience fatigue, headache, fever, and joint and muscle pains. The saliva from lone star ticks can be irritating, however, redness and discomfort at a bite site do not necessarily indicate an infection.

Unlike *Borrelia burgdorferi*, STARI has not been linked to arthritis, neurologic disease, or chronic symptoms.

Diagnosis

STARI is diagnosed on the basis of symptoms, geographic location, and possibility of tick bite. Since the cause of STARI is unknown, no diagnostic blood tests have been developed. Researchers once hypothesized that STARI was caused by the spirochete, *Borrelia lonestari*, however, further research did not support this idea and the cause of STARI remains unknown.

Treatment

It is not known whether antibiotic treatment is necessary or beneficial. Nonetheless, because it resembles early LD, STARI is treated with oral antibiotics.

Preventive measures

Ticks can spread to other organisms that may cause a different type of rash. People should monitor their health closely after any tick bite, and should consult their physician if they experience a rash, fever, headache, joint and muscle pains, or swollen lymph nodes within 30 days of a tick bite. These can be signs of a number of tick-borne diseases.

Tick-borne illness may be prevented by avoiding tick habitats (dense woods and brushy areas), using insect repellents containing DEET or permethrin, wearing long pants and socks, performing tick checks, and promptly removing ticks after outdoor activity. Additional prevention types are available.

Distinctions between STARI and LD symptoms

In a study that compared physical findings from STARI patients in Missouri with LD patients in New York, several key differences were noted:

- Patients with STARI were more likely to recall a tick bite than were patients with LD;

- The time period from tick bite to onset of the skin lesion was shorter among patients with STARI (6 days, on average);

- STARI patients with an *erythema migrans* rash were less likely to have other symptoms than were LD patients with that rash;

- STARI patients were less likely to have multiple skin lesions, had lesions that were smaller in size than LD patients (6-10 cm for STARI *versus* 6-28 cm for LD), and had lesions that were more circular in shape and with more central clearing; and

- After antibiotic treatment, STARI patients recovered more rapidly than did LD patients.

Take-home points

- There is more to say about rashes without them being LD. STARI, a rash similar to the LD rash, has been described in humans following bites of the lone star tick, *Amblyomma americanum*. The adult female is distinguished by a white dot or "lone star" on her back. All three life stages (lymph, larva, adult) of *Amblyomma americanum* aggressively bite people.

- Lone star tick does not cause LD. It may be accompanied by fatigue, fever, headache, muscle

Lyme disease

and joint pains, and will occasionally develop a circular rash similar to the rash of early LD. The cause of this rash has not been determined. Symptoms resolve following treatment with an oral antibiotic (doxycycline), but it is unknown whether this medication only sped recovery. Lone star ticks do not transmit *Borrelia burgdorferi*. This tick's saliva can be irritating with redness and discomfort at a bite site, does not necessarily indicate an infection, and has in fact been shown to kill *Borrelia*.

- In the U.S., the lone star tick is found throughout the eastern, southeastern, and south-central States. Its distribution, range, and abundance have increased over the past 20-30 years. All three life stages (larva, nymph, adult) will feed on humans and domesticated animals, and may be quite aggressive.

- STARI is diagnosed on the basis of symptoms, geographic location, and possibility of tick bite. Since the cause of STARI is unknown, no diagnostic blood tests have been developed. It is not known whether antibiotic treatment is necessary or beneficial. Nonetheless, because it resembles early LD, STARI is treated with oral antibiotics.

- STARI and other tick-borne illness may be prevented by avoiding tick habitats (dense woods and brushy areas), using insect repellents containing DEET or permethrin, wearing long pants and socks, performing tick checks, and promptly removing ticks after outdoor activity.

12

What you need to know about Lyme neuroborreliosis

What is LNB?

Lyme *neuroborrelliosis* (LNB) is a disorder of the central nervous system (CNS). A neurological manifestation of Lyme, LNB is caused by a systemic infection of spirochetes of the genus *Borrelia*.

Signs and symptoms

Neuroborreliosis is often preceded by the typical symptoms of LD, which include the *erythema migrans* (EM) rash and flu-like symptoms such as fatigue, fever, headache, muscle and joint pains. Its neurologic symptoms include meningo-radiculitis (more common in European patients), cranial nerve abnormalities, and altered mental status. Sensory findings may also be present. Rarely, a progressive form of encephalomyelitis may occur (see next Chapter 13). In children, symptoms of *neuroborreliosis* include headache, sleep disturbance, and symptoms associated with increased intracranial pressure (such as papilledema) can occur. Less common childhood symptoms can include meningitis, myelitis ataxia, and chorea. Ocular LD has also been reported, as has *neuroborreliosis* affecting the spinal cord, but neither of these findings are common.

Differential diagnosis

A number of diseases can produce symptoms similar to those of LNB. They are discussed separately in accompanying Sidebars and include:

- **Alzheimer's disease (AD);**

- **Acute disseminated encephalomyelitis (ADE);**

- **Viral meningitis (VM);**

- **Multiple sclerosis (MS);**

- **Bell's palsy (BP);** and

- **Amyotrophic lateral sclerosis (ALS)** (aka **Lou Gehring disease).**

Diagnosis is determined by clinical examination of visible symptoms. *Neuroborreliosis* can also be diagnosed serologically to confirm clinical examination via western blot, ELISA, and PCR (see Chapter 5 for definitions and explanations of these assays).

Treatment

In the U.S., *neuroborreliosis* is typically treated with intravenous antibiotics which cross the blood-brain barrier (BBB) such as penicillin, ceftriaxone, or cefotaxime (see Table 7.1). One relatively small randomized clinical trial (RCT) suggested that ceftriaxone was more effective than penicillin in the treatment of LNB. Other small observational studies suggested ceftriaxone is also effective in children. The recommended duration of treatment is 2-4 weeks.

Several European studies have suggested that oral doxycycline is equally as effective as intravenous ceftriaxone in treating *neuroborreliosis*. Doxycycline has not been widely studied as a treatment in the U.S., but antibiotic sensitivities of prevailing European and U.S. isolates of *Borrelia burgdorferi* tend to be identical. However, doxycycline is generally not prescribed to children due to the risk of bone and tooth damage.

There are several discredited or doubtful treatments for *neuroborreliosis,* including:

- *Malaria therapy;*

- *Hyperbaric oxygen therapy;*

- *Colloidal silver therapy; and*

- *Injections of hydrogen peroxide and mismacine.*

Take-home points

- Lyme neuroborrelliosis, a manifestation of Lyme, is a disorder of the central nervous system.

- *Neuroborreliosis* is caused by a systemic infection of spirochetes of the genus *Borrelia.*

- Often preceded by the typical symptoms of Lyme disease (*erythema migrans* rash, flu-like symptoms, etc.), *neuroborreliosis* symptoms are physical, neurological (meningo-radiculitis, cranial nerve abnormalities, and altered mental status), and sensory (encephalomyelitis).

- In children, common symptoms of *neuroborreliosis* include headache, sleep disturbance, and

Lyme disease

intracranial pressure such as papilledema. Less common symptoms include meningitis, myelitis ataxia, and chorea.

- The diagnosis of *neuroborreliosis* is determined differentially by clinical examination of visible symptoms and serologically. Confounding other diseases include: Alzheimer's disease; acute disseminated encephalomyelitis; viral meningitis; multiple sclerosis; Bell's palsy; and amyotrophic lateral sclerosis that are further separately discussed in following Sidebars.

- Neuroborreliosis is typically treated with intravenous antibiotics which cross the blood-brain barrier such as penicillin, ceftriaxone, or cefotaxime.

- There are several discredited or doubtful treatments for *neuroborreliosis* including: malaria therapy; hyperbaric oxygen therapy; colloidal silver, and injections of hydrogen peroxide and mismacine.

Sidebar 12.1
Alzheimer's disease

According to current knowledge, Alzheimer's disease (AD) is an age-related (not age-caused), progressive, purportedly irreversible chronic neurological disorder in which brain cells die slowly destroying memory and thinking skills, eventually even the ability to carry out the simplest tasks. In most people with AD, symptoms first appear after age 60. The memory loss and the associated cognitive decline had developed for a period of years or decades. When it appears in persons aged in their 30s and mid-60s, it is called *early-onset AD* but the more prevalent form occurs in the mid-60s and older and is called *late-onset AD*. In the absence of disease, the human brain often can function well into the 10th decade of life.

In healthy people, all sensations, movements, thoughts, memories, and feelings are the result of signals that pass through billions of nerve cells (or neurons) in the brain. Neurons constantly communicate with each other through electrical charges that travel down axons, causing the release of chemicals across tiny gaps to neighboring neurons. Other cells in the brain, such as astrocytes and microglia (the brain's "cleaning maids") clear away debris and help keep neurons healthy. In a person with AD, toxic changes in the brain destroy this healthy balance over years, even decades, Researchers believe that this process involves two proteins called "amyloid-beta" and "tau", which somehow become toxic to the brain. It appears that abnormal tau accumulates, eventually forming tangles inside neurons. Also, amyloid-beta clumps into plaques, which slowly build up between neurons. As the level of amyloid reaches a tipping point, there is a rapid spread of tau throughout the brain, pointing to interactions between these two types of proteins. Such interactions have been found to be important and are currently being studied.

But, tau and amyloid-beta proteins may not be the only factors involved in Alzheimer's. Other changes that affect the brain may also play a role over time. Thus, the vascular system may fail to deliver sufficient blood and nutrients to the brain; the brain may lack the glucose needed to power its activity; chronic inflammation sets in as microglial cells fail to clear away debris; and astrocytes react to distressed microglia. Eventually, neurons lose their ability to communicate and, as they die, the brain shrinks. This decay process begins in the hippocampus, that part of the brain that is important to learning and memory (see Figure S12.1.1). People may begin to experience memory loss, impaired decision-making, and language problems. As more neurons die throughout the brain, a person with Alzheimer's gradually loses the ability to think, remember, make decisions, and function independently.

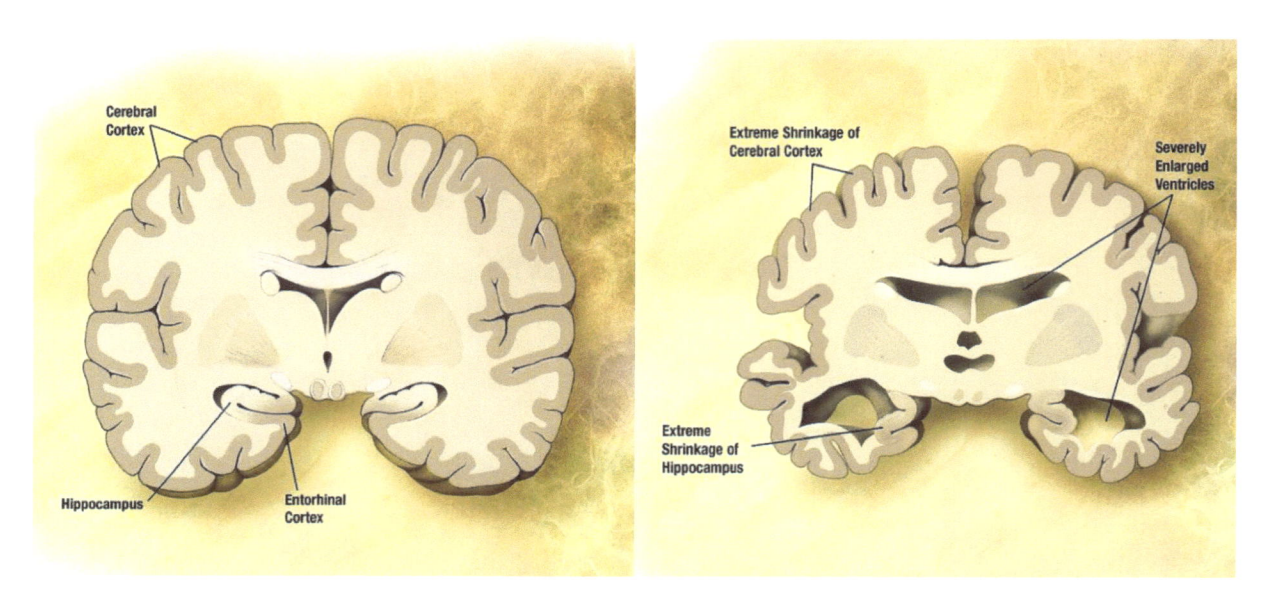

Figure S12.1.1 – Contrasting a healthy brain with a severe Alzheimer-diseased brain

It is readily seen in Figure S12.1.1 that the brain structure and convolutions are distorted with extreme shrinkage of the cerebral cortex and the hippocampus and severely enlarged ventricles. The extent of brain shrinkage may be used as a rough gauge for assessing the severity of the disease.

Achieving a deeper understanding of the molecular and cellular mechanisms—and how they may interact—is vital to the development of effective therapies. Much progress has been made in identifying various underlying factors. Additionally, advances in brain imaging allow us to see the course of plaques and tangles in the living brain. Blood and fluid biomarkers are further providing insights about when the disease starts and how it progresses. More is also known about the genetic underpinnings of the disease and how they can affect particular biological pathways. These advances enable the development and testing of promising new therapies, including: drugs that reduce or clear the increase of tau and amyloid proteins in the brain; therapies that target the vascular system, glucose metabolism, and inflammation; lifestyle interventions, like exercise and diet; and behavioral approaches like social engagement and integration that may enhance brain health. Research is moving quickly, ever closer to the day when we can delay or even prevent the devastation of AD and dementia.

Figure S12.1.2 – Alzheimer's disease – Demystifying the disease and what you can do about it

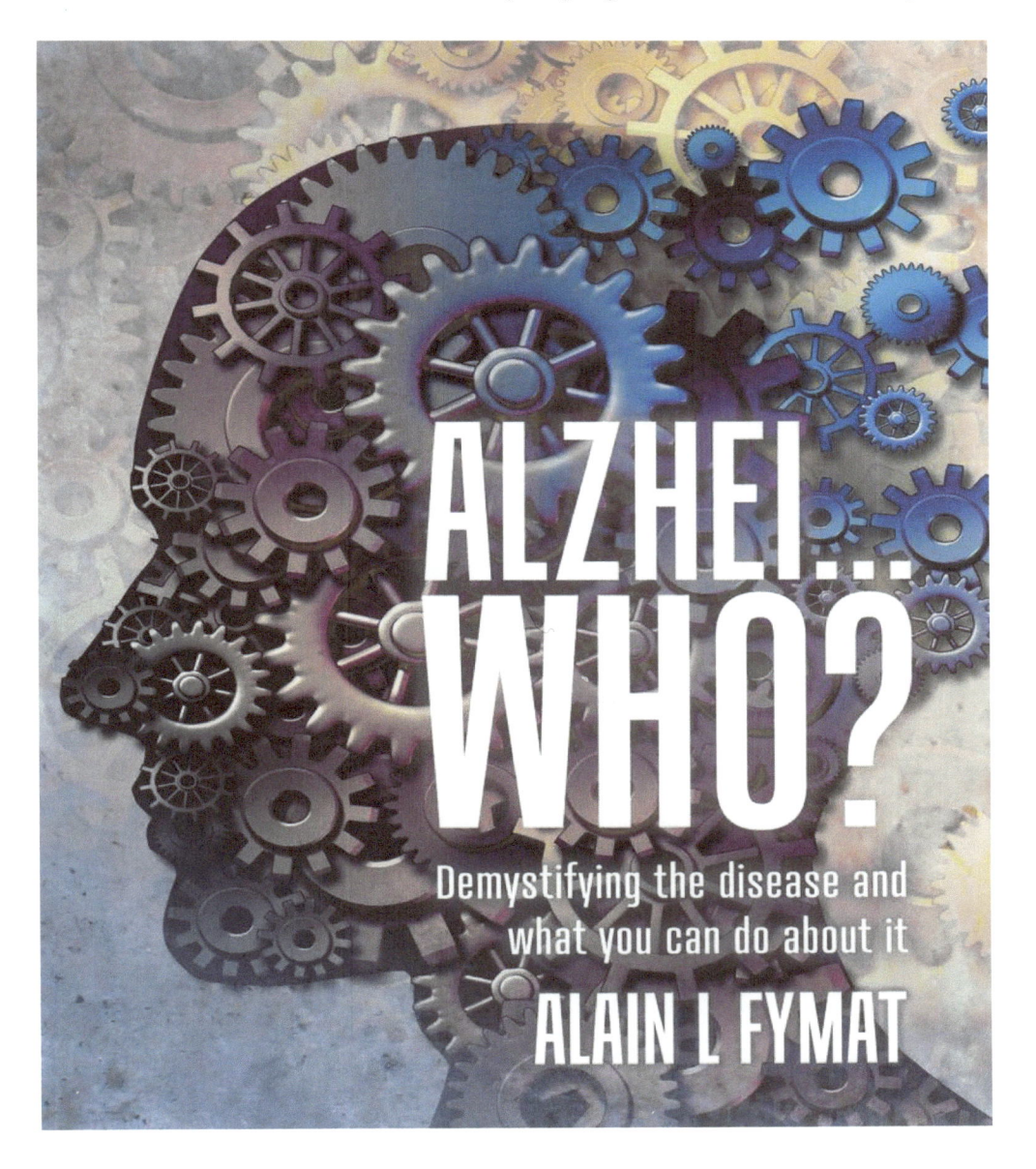

For those readers interested in knowing more about Alzheimer's disease and its relationship to Lyme disease, I refer them to my book on the subject (Figure S12.1.2).

Sidebar 12.2
Acute disseminated encephalomyelitis

Lyme disease

Acute disseminated encephalomyelitis (ADEM), or acute demyelinating encephalomyelitis (ADE), is a rare autoimmune disease marked by a sudden, widespread attack of inflammation in the brain and spinal cord. As well as causing the brain and spinal cord to become inflamed, ADEM also attacks the nerves of the central nervous system (CNS) and damages their myelin insulation, which, as a result, destroys the white matter. It is often triggered by a viral infection or, perhaps exceedingly rarely, specific non-routine vaccinations.

ADEM's symptoms resemble the symptoms of multiple sclerosis (MS, see Sidebar 12.4), so the disease itself is sorted into the classification of the multiple sclerosis borderline diseases. However, it has several features that distinguish it from MS in that it occurs usually in children and is marked with rapid fever, although adolescents and adults can get the disease too. ADEM consists of a single flare-up whereas MS is marked with several flare-ups (or relapses), over a long period of time. Relapses following ADEM are reported in up to a quarter of patients, but the majority of these "multiphasic" presentations following ADEM likely represent MS. ADEM is also distinguished by a loss of consciousness, coma, and death, which is very rare in MS, except in severe cases.

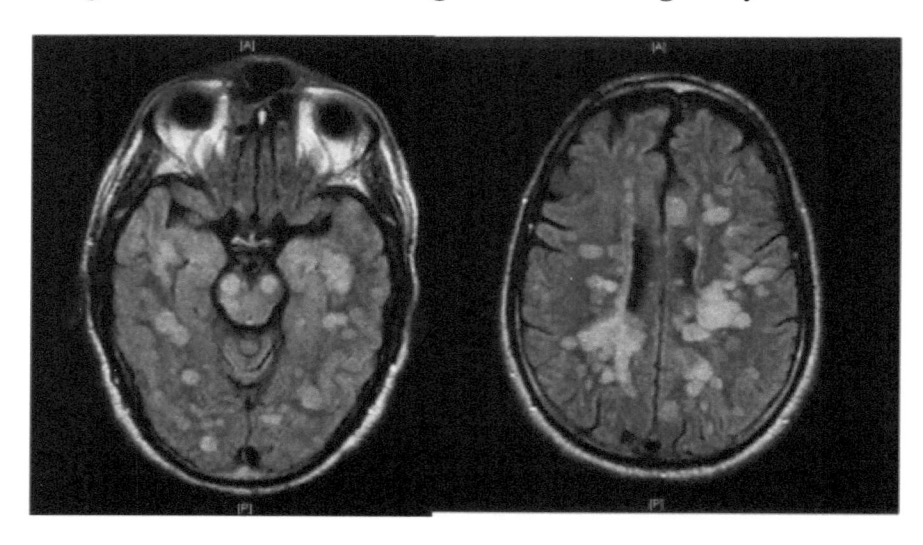

Figure S12.2.1 - Fulminating ADEM showing many lesions

Reference: Rodríguez-Porcel F, Hornik A, Rosenblum J and Biller EB
https://openi.nlm.nih.gov/detailedresult.php?img=PMC4274983_fneur-05-00270-
g002&query=acute+disseminated+encephalomyelitis&lic=by&req=4&npos=17

ADEM affects about 8 per 1,000,000 people per year. Although it occurs at all ages, most reported cases are in children and adolescents with average age around 5 to 8 years old. The disease affects males and females almost equally. It shows seasonal variation with higher incidence in winter and spring months, which may coincide with higher viral infections during these months. The mortality rate may be as high as 5%; however, full recovery is seen in 50%-75% of cases with increase in survival rates up to 70%-90%, including minor residual disability as well. The average time to recover from ADEM flare-ups is one to six months.

ADEM produces multiple inflammatory lesions in the brain and spinal cord, particularly in the white

Lyme disease

matter. Usually these are found in the subcortical and central white matter and cortical gray-white junction of both cerebral hemispheres, brainstem, and spinal cord, but periventricular white matter and gray matter of the cortex, thalamus, and basal ganglia may also be involved.

When a person has more than one demyelinating episode of ADEM, the disease is then called recurrent disseminated encephalomyelitis (RDEM) or multiphasic disseminated encephalomyelitis (MDEM). Also, a fulminant course in adults has been described as shown in Figure S12.2.1. In that illustrated case, the patient survived, but remained in a persistent vegetative state.

Sidebar 12.3
Viral meningitis

Viral meningitis (VM), also known as aseptic meningitis (AM), is a type of meningitis due to a viral infection. It results in inflammation of the meninges (the membranes covering the brain and spinal cord). Symptoms commonly include headache, fever, sensitivity to light, and neck stiffness.

Figure S12.3.1 – Depicting the meninges of the central nervous system

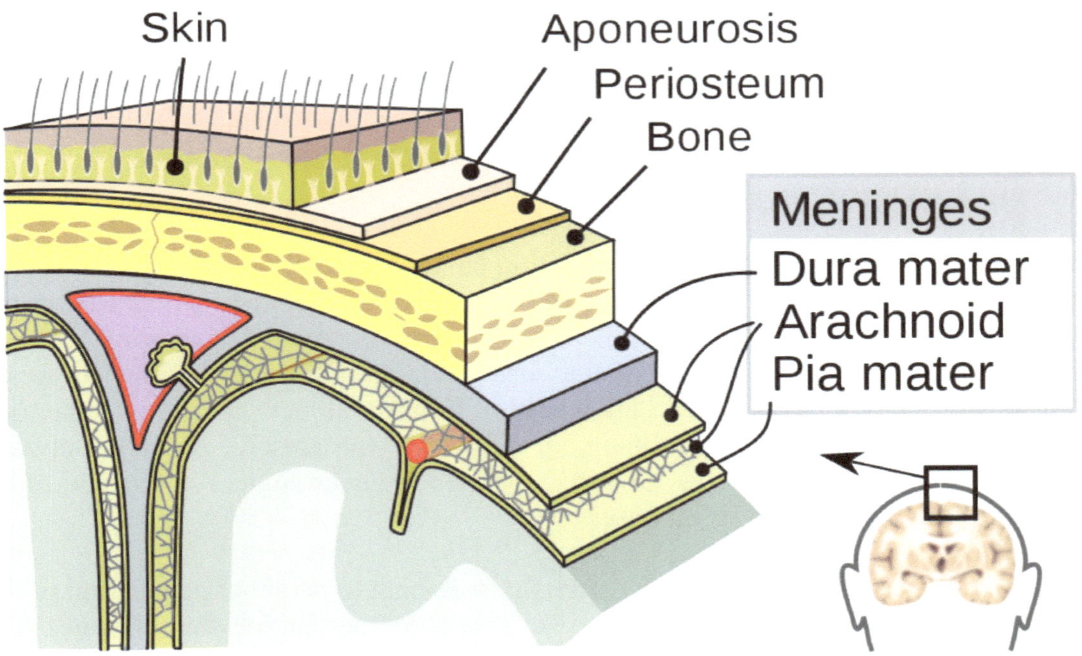

Reference: SVG by Mysid, original by SEER Development Team

Viruses (see Chapter 14) are the most common cause of AM. Most cases of VM are caused by enteroviruses (common stomach viruses). However, other viruses can also cause viral meningitis such as, for instance, the West Nile, mumps, measles, herpes simplex types (HSV) I and II, varicella, and lymphocytic choriomeningitis (LCM) viruses. Based on clinical symptoms, VM cannot be reliably differentiated from bacterial meningitis (BM), although VM typically follows a more benign clinical course. VM has no evidence of bacteria present in the CSF. Therefore, lumbar puncture with CSF analysis is often needed to identify the disease.

In most cases, there is no specific treatment, with efforts generally aimed at relieving symptoms (headache, fever or nausea). A few viral causes, such as HSV, have specific treatments.

In the United States, VM is the cause of more than half of all cases of meningitis. With the prevalence of BM in decline, the viral disease is garnering more and more attention. The estimated incidence has a considerable range, from 0.26 to 17 cases per 100,000 people. For enteroviral meningitis (EM), the most common cause of VM, there are up to 75,000 cases annually in the U.S. alone. While the disease can occur in both children and adults, it is more common in children.

Figure S12.3.1 is a depiction of the several meninges of the central nervous system.

Sidebar 12.4
Multiple sclerosis

Multiple sclerosis (MS) was first described in 1868 by the French neurologist Jean-Marie Charcot. The name refers to the numerous glial scars (or *sclerae* – essentially plaques or lesions) that develop on the white matter of the brain and spinal cord. Figure S12.4.1 is a photomicrograph of a demyelinating MS-lesion obtained by immunohistochemical staining for CD68 (original magnification 10x. Marvin 101). It highlights numerous macrophages (in brown color).

MS is a demyelinating disease in which the insulating covers of the nerve cells in the brain and spinal cord are damaged. This damage disrupts the ability of parts of the nervous system to transmit signals, resulting in a range of signs and symptoms, including physical, mental, and sometimes psychiatric problems. Specific symptoms can include double vision, blindness in one eye, muscle weakness, and trouble with sensation or coordination. MS takes several forms, with new symptoms either occurring in isolated attacks (relapsing forms) or building up over time (progressive forms). Between attacks, symptoms may disappear completely; however, permanent neurological problems often remain, especially with the advancement of the disease.

While the cause is unclear, the underlying mechanism is thought to be either destruction of the immune system or failure of the myelin-producing cells. Proposed causes for this include genetics and environmental factors such as being triggered by a viral infection. MS is usually diagnosed based on the presenting signs and symptoms and the results of supporting medical tests.

There is no known cure for MS. Treatments attempt to improve function after an attack and prevent new attacks. While modestly effective, medications used can have side effects and be poorly tolerated. Physical therapy can help with people's ability to function. Many people pursue alternative treatments.

Lyme disease

despite a lack of evidence of benefit. The long-term outcome is difficult to predict, with good outcomes more often seen in women, particularly those who develop the disease early in life, those with a relapsing course, and those who initially experienced few attacks. Life expectancy is on average 5 to 10 years lower than that of the unaffected population. A number of new treatments and diagnostic methods are under development.

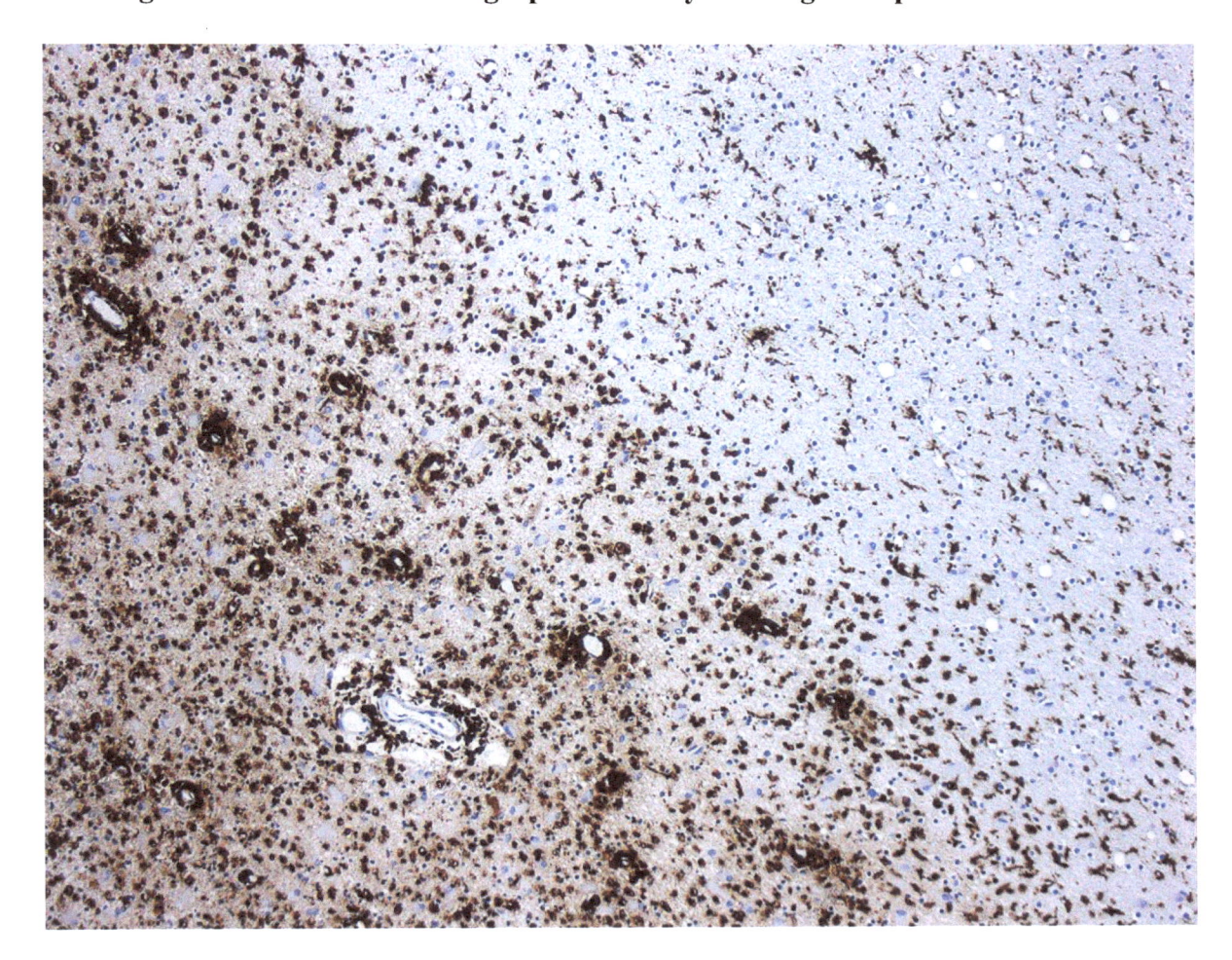

Figure S12.4.1 - Photomicrograph of a demyelinating multiple sclerosis lesion

MS is the most common immune-mediated disorder affecting the central nervous system. In 2015, about 2.3 million people were affected globally, with rates varying widely in different regions and among different populations. In that year, about 18,900 people died, up from 12,000 in 1990. The disease usually begins between the ages of 20 and 50 and is twice as common in women as in men.

Sidebar 12.5
Bell's palsy

Bell's Palsy (BP) is a type of facial paralysis that results in an inability to control the facial muscles on the affected side (Figure S12.5.1). It is named after Scottish surgeon Charles Bell (1774–1842), who first described the connection of the facial nerve to the condition.

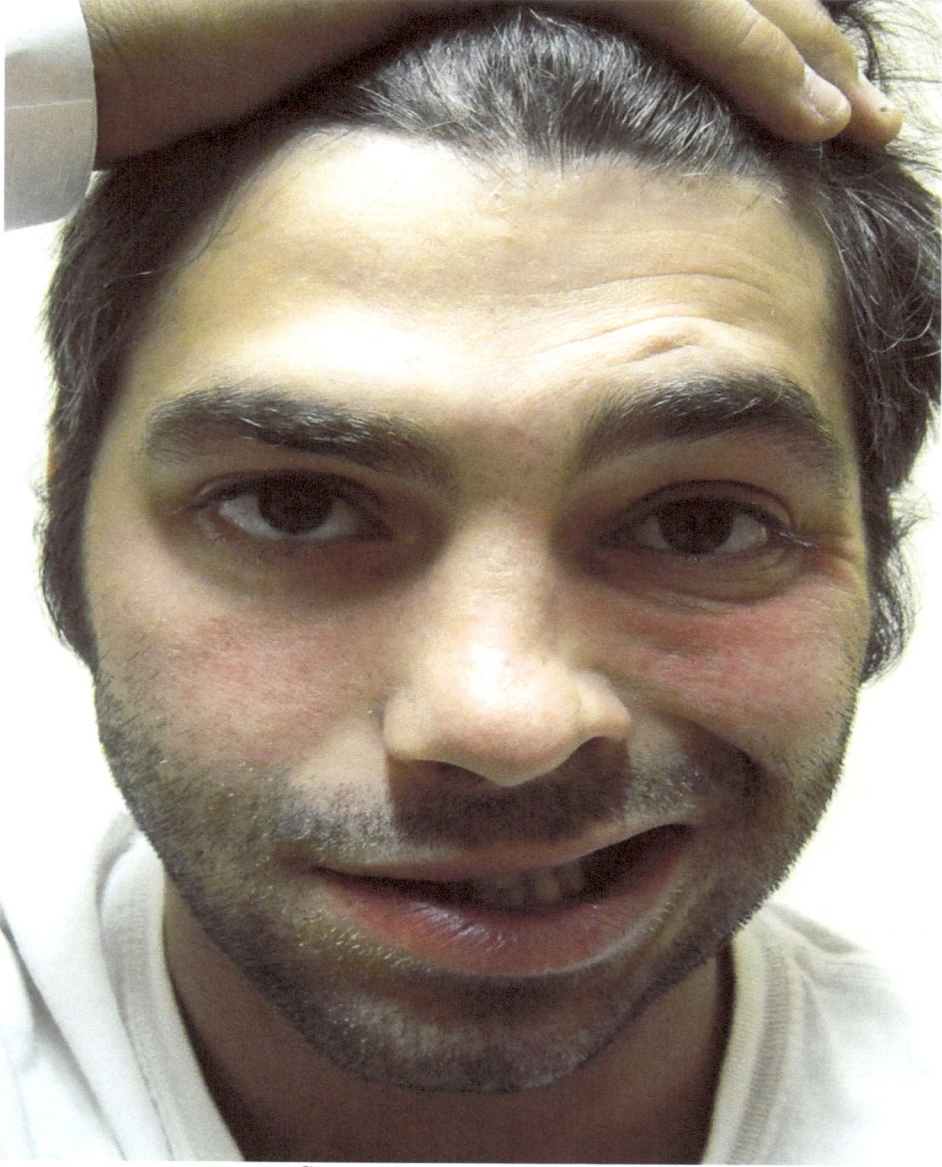

Figure S12.5.1 – Bell's palsy

Source: James Wellman, MD

Symptoms can vary from mild to severe. They may include muscle twitching, weakness, or total loss of the ability to move one or rarely both sides of the face. Other symptoms include drooping of the eyelid, a change in taste, pain around the ear, and increased sensitivity to sound. Typically symptoms come on over 48 hours.

The cause of Bell's palsy is unknown. Risk factors include diabetes, a recent respiratory tract infection, and pregnancy. It results from a dysfunction of cranial nerve VII (the facial nerve). Many believe that this is due to a viral infection that results in swelling. Diagnosis is based on a person's appearance and ruling out other possible causes. Other conditions that can cause facial weakness include brain tumor, stroke, Ramsay-Hunt syndrome (RHS), myasthena gravis (MG), and Lyme disease.

The condition normally gets better by itself with most patients achieving normal or near-normal function. Corticosteroids have been found to improve outcomes, while antiviral medications may be of a small additional benefit. The eye should be protected from drying-up with the use of eye drops or an eyepatch. Surgery is generally not recommended. Often signs of improvement begin within 14 days, with complete recovery within six months. A few may not recover completely or have a recurrence of symptoms.

Bell's palsy is the most common cause of one-sided facial nerve paralysis (70%). It occurs in 1%-4% per 10,000 people per year. About 1.5% of people are affected at some point in their life. It most commonly occurs in people between ages 15 and 60. Males and females are affected equally.

Sidebar 12.6
Amyotrophic lateral sclerosis

Amyotrophic lateral sclerosis (ALS), also known as motor neuron disease (MND) or Lou Gehrig's disease, is a disease that causes the death of neurons controlling voluntary muscles. MND is also the name for a group of conditions of which ALS is the most common.

Descriptions of the disease date back to at least 1824 by Charles Bell (of the famed Bell's palsy reviewed in Sidebar S12.5). In 1869, the connection between the symptoms and the underlying neurological problems was first described by Jean-Marie Charcot who, in 1874, began using the term *amyotrophic lateral sclerosis*. It became well-known in the United States in the 20th century when, in 1939, it affected baseball player Lou Gehrig, and later worldwide following the 1963 diagnosis of cosmologist Stephen Hawking. The first ALS gene was discovered in 1993 while the first animal model was developed in 1994. In 2014, videos of the Ice Bucket Challenge went viral on the Internet and increased public awareness of the condition.

ALS is characterized by stiff muscles, muscle twitching, and gradually worsening weakness due to muscles decreasing in size. It may begin with weakness in the arms or legs, or with difficulty speaking or swallowing. About half of the people affected develop at least mild difficulties with thinking and behavior and most people experience pain. Most patients eventually lose the ability to walk, use their hands, speak, swallow, and breathe.

Lyme disease

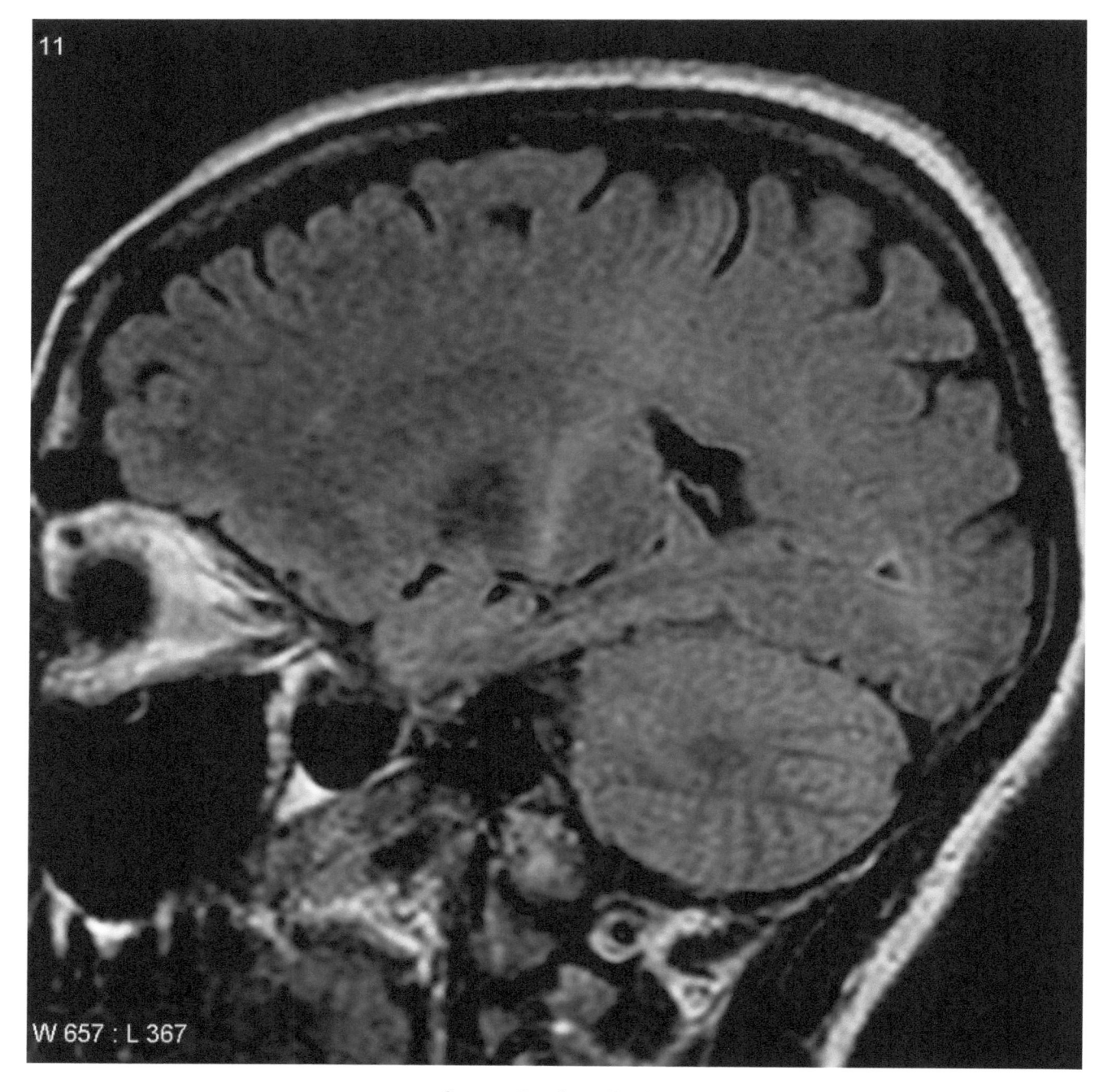

Figure S12.6.1 – Amyotrophic lateral sclerosis

Source: Frank Gaillard
(http://radiopaedia.org/uploads/radio/0001/2754/ALS_Coronal.jpg)

The cause is not known in 90%-95% of cases, but is believed to involve both genetic and environmental factors. The remaining 5%–10% of cases are inherited from a person's parents. About half of these genetic cases are due to one of two specific genes. The underlying mechanism involves damage to both upper and lower motor neurons. The diagnosis is based on a person's signs and symptoms, with testing done to rule out other potential causes.

No cure for ALS is known. The goal of treatment is to improve symptoms. A medication called *Riluzole* may extend life by about 2-3 months. Non-invasive ventilation may result in both improved quality and length of life. Mechanical ventilation can prolong survival but does not stop disease progression. A feeding tube may help. The disease can affect people of any age, but usually starts around the age of 60 and, in inherited cases, around the age of 50. The average survival from onset to death is 2-4 years, although this can vary. About 10% survive longer than 10 years. Most die from respiratory failure.

In Europe, the disease affects about 2-3 people per 100,000 per year. Rates in much of the world are unclear. In the United States, it is more common in white people than black people.

13

What you need to know about Lyme carditis

Lyme disease is caused by the spread of bacterial infection involving the *Borrelia burgdorferi sensu lato* (Bbsl) genus. The pathology and symptoms of Lyme disease condition correspond to the progressive involvement of multiple organs and systems in the body. Thus, evidence of bacterial infection has been found in the skin, heart, peripheral and central nervous system, and the joints. In some cases, the presence of bacteria in organs such as the heart has only been found after autopsy and may have played a role in the cause of a patient's death due to long-term damage going undetected.

What is Lyme carditis?

Lyme carditis (LC) is an uncommon manifestation of early-disseminated LD. It is a disease of the heart. Early signs of heart problems are not often associated with an infection such as *borreliosis* and so the condition can easily be overlooked. Carditis means inflammation of the heart, in comparison to cardiomegaly, which is enlargement of the heart. The former condition is generally acute whereas the latter usually takes some time to develop and is often a result of a circulatory issue causing the heart to grow in an effort to compensate for dysfunction elsewhere.

LC occurs when Lyme disease bacteria enter the tissues of the heart. This can interfere with the normal movement of electrical signals from the heart's upper to lower chambers, a process that coordinates the beating of the heart. The result is something physicians call "heart block", which can be mild, moderate, or severe. Heart block from LC can progress rapidly. LC is a heart problem that is usually fully treatable and leaves no lasting damage if caught early enough. However, long-standing inflammation in the heart can cause permanent tissue damage and ongoing heart problems may result from LD whether treated or untreated.

Other forms of presentation of LC include myocarditis, endocarditis, valvular heart disease, pericarditis, and myopericarditis. However, such conditions have been reported less frequently than the typical conduction system abnormalities.

How do you get heart disease with LD infection?

It is not necessarily the case that those with an existing heart condition will succumb to LC. Lyme disease heart problems may have no prior cardiac dysfunction and no persistent problems after treatment for the infection. However, there is a danger that those who do already suffer from symptoms of heart disease (such as shortness of breath, chest pains, fatigue, and palpitations), could simply ignore these early signs of LC or that their physician may dismiss them as part of their existing heart condition. Any symptom exacerbation, or the occurrence of new symptoms, should be brought to the attention of a qualified physician for assessment.

… and what are the symptoms?

As well as those symptoms of heart disease mentioned above, LC symptoms can include some or many of the following: breathing difficulties when lying down or sleeping, chest pain, dizziness, dyspnea, faintness, heart palpitations, light-headedness, myalgia, shortness of breath, throbbing in the neck, and syncope. They usually have other symptoms as well such as fever and body aches, and they may have more specific symptoms of LD, such as the *erythema migrans* rash. Feverishness and other signs of infection may or may not be present and patients may recall a recent bout of a flu-like illness, a LD rash, and also be experiencing other symptoms of LD such as arthritis in one or more large joints, such as the knees. In sum, the symptoms of LC are caused by the bacterial infection compromising the electrical signaling in the heart, creating an abnormal heart rate which shows up on an electrocardiogram. Early treatment of LD is important and can prevent complications such as LC.

How common is LC?

Based on (U.S.) national surveillance data from 2001-2010, LC occurred in approximately 1% of LD cases reported to the CDC&P. It is important to recognize its three "early" factors:

> *Early recognition:* Atrioventricular (AV) block caused by LC can occur as early as three days after a tick bite;

> *Early consideration:* LC should be considered in patients with AV block, especially those living in an endemic area; and

> *Early treatment:* Early recognition and appropriate antibiotic treatment are essential to achieve a favorable prognosis.

Testing for LD heart problems

An electrocardiogram (EKG) is the test usually performed to assess the electrical signaling of the heart and the regulation of the heart rate. It may show an atrioventricular (AV) heart block, which means that the four chambers of the heart are not communicating correctly in order to manage the valves and the emptying and filling of the sections of the heart. The degree to which this pump process is disrupted can vary enormously between patients and some estimate that one in ten patients with LD has some sort of cardiovascular (CV) disruption due to the infection. However, many are undiagnosed as many physicians lack experience in recognizing the symptoms, particularly as heart disease is common in the general population. Electrophysiologists are best educated and trained for this task. Other conditions such as lupus may also result in symptoms such as AV block and can further confuse diagnosis.

Figure 13.1 is a sample EKG showing an abnormal heart rhythm that can indicate Lyme carditis.

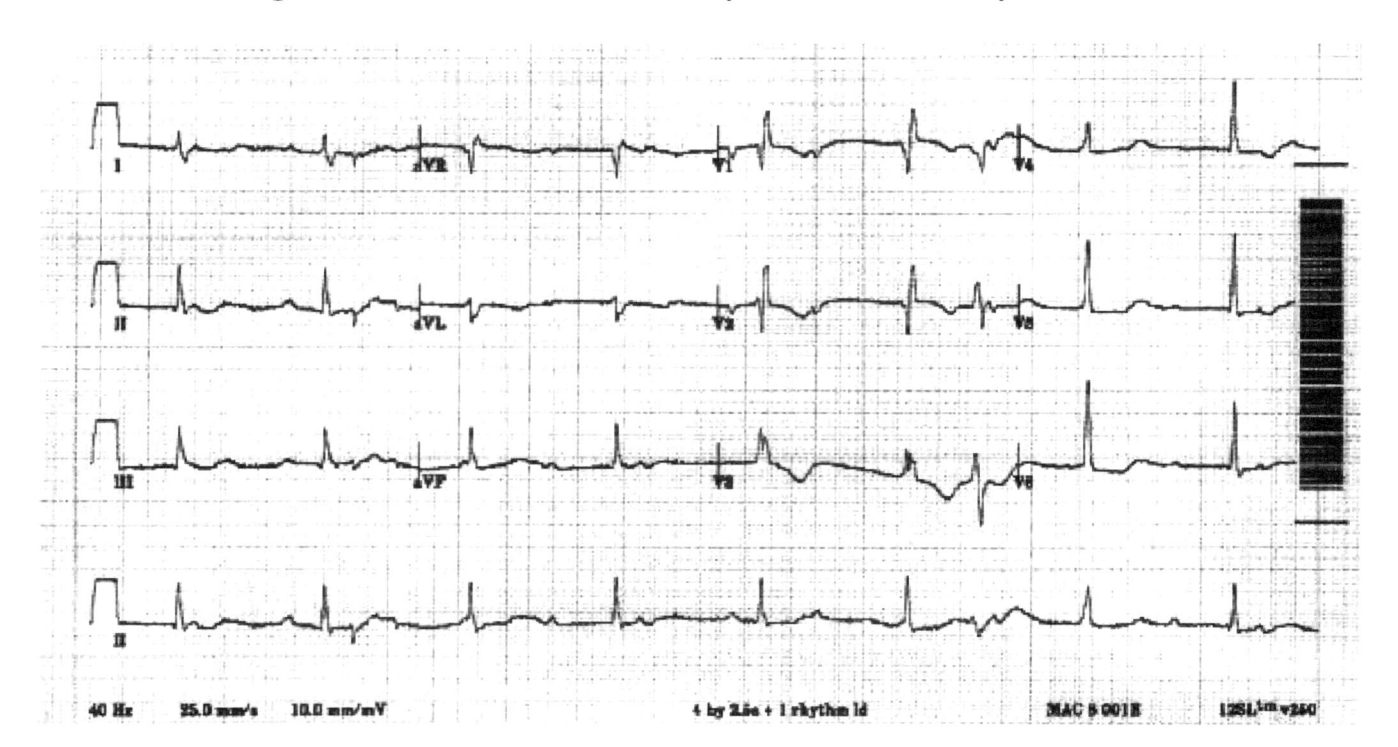

Figure 13.1 - An abnormal heart rhythm can indicate Lyme carditis

How is LC treated?

The treatment is a two-to-four week course of an antibiotic such as doxycycline or ceftriaxone. Serious cases of AV heart block can necessitate hospitalization, IV antibiotics, and even a pacemaker to provide temporary assistance in regulating the heart as the infection is fought. Around a third of patients with LC require such a pacemaker although almost all of these have no need of the medical device after the LD infection has been eradicated. As many LD infections are thought to remain unrecognized and, therefore, often untreated, it is possible that fatal cases of LC occur without anyone realizing, particularly in those with pre-existing heart disease. Autopsies may reveal the presence of bacteria in the heart tissue but it is difficult to surmise the contribution of such bacterial infection to the cause of death in some patients and so LD deaths from heart disease may be under-reported. However, with trained and skillful medical personnel, it is possible to prevent a fatal conclusion.

The importance of timely diagnosis and appropriate management to achieve a favorable prognosis cannot be over-emphasized.

Can LC be fatal?

Yes, between 1985 and 2019, eleven cases of fatal LC have been reported in the medical literature.

How long does it take for a person to recover from LC?

Typically the patient receives antibiotic treatment for 14-21 days. Most symptoms are gone within 1-6 weeks.

Key points for healthcare providers

- Ask all patients with suspected LD about cardiac symptoms, e.g., chest pain, difficulty breathing with exertion, fainting, light headedness, palpitations, and shortness of breath;

- Ask patients with acute, unexplained cardiac symptoms about possible tick exposure and symptoms of LD;

- Treat patients with suspected LC with appropriate antibiotics immediately – do not wait for LD test results; and

- Talk to patients about tick bite prevention.

Take-home points

- Lyme carditis is an uncommon manifestation of early-disseminated Lyme disease. It is a disease of the heart. It is usually fully treatable and leaves no lasting damage if caught early enough. However, long-standing inflammation in the heart can cause permanent tissue damage and ongoing heart problems may result from Lyme disease whether treated or untreated.

- Lyme carditis occurs when Lyme disease bacteria enter the tissues of the heart, interfering with the normal movement of electrical signals from the heart's upper to lower chambers, and resulting in a "heart block".

- A heart block from Lyme carditis can progress rapidly.

- It is not necessarily the case that those with an existing heart condition will succumb to Lyme carditis, however, their treating physician may not identify the condition.

- Symptoms of Lyme carditis can include: breathing difficulties when lying down or sleeping, chest pain, dizziness, dyspnea, faintness, heart palpitations, light-headedness, myalgia, shortness of breath, throbbing in the neck, and syncope.

- During 2001-2010, Lyme carditis occurred in approximately 1% of Lyme cases reported. Early recognition, early consideration, and early treatment are essential to achieve a favorable prognosis.

- An electrocardiogram performed to assess the electrical signaling of the heart and the regulation of the heart rate may show an atrioventricular heart block, which means that the four chambers of the heart are not communicating correctly in order to manage the valves and the emptying and filling of the sections of the heart.

- Electrophysiologists are best educated and trained for the task of diagnosing Lyme carditis. The importance of timely diagnosis and appropriate management to achieve a favorable prognosis cannot be over-emphasized..

- Lyme carditis is usually treated by a two-to-four week course of an antibiotic (doxycycline or ceftriaxone), serious cases of atrioventricular heart block necessitating hospitalization, IV antibiotics, and even the placement of a temporary pacemaker. Most symptoms are gone within 1-6 weeks.

14

Pathogens...in the brain?

Cytomegalovirus (CMV) is one of the co-infections of LD that are most frequently cited. However, CMV is but one of several brain pathogens. While not all such pathogens have yet been recognized in Lyme *neuroborreliosis*, I will summarize our state-of-knowledge of them beginning with CMV.

Cytomegalovirus

Cytomegalovirus is a CLD co-infection and also a cancer-causing agent. Approximately 80% of adults in the U.S. and most people across the globe are infected with it. CMV is a member of the herpes family, which I will discuss later. It shares with it the common ability to remain alive, yet dormant, in the human body for the life of its human host. Rarely does CMV become active unless the immune system is weakened and rendered unable to hold the virus in check. Trauma or a severely impaired immune system with complex conditions such as LD could awaken CMV. Table 14.1 is a convenient summary of our current knowledge of CMV.

Table 14.1 – State-of-knowledge of the cytomegalovirus

Signs and symptoms	CASE OF INTACT IMMUNE SYSTEM: o Chills o Fatigue: severe o Fever: prolonged, high o Headache o Ill: overall feeling o Lymph glands: swollen o Spleen: enlarged
	CASE OF WEAKENED IMMUNE SYSTEM: o Anemia o Blindness o Death (in some cases) o Liver infections o Pneumonia
	PATIENTS MOST AT RISK: Persons with: o Chronic infections: such as CLD

Lyme disease

	o HIV infections o Zoonotic infections o Transplanted organ(s)
Virus transmission	o Person-to-person through direct contact or/and bodily fluids exchange (breast milk, blood, cervical secretions, semen, urine) o Passed from pregnant woman to fetus
Treatment	o Re-establish a strong immune system o Eliminate infections that compromise the immune system

Other brain pathogens

Bacteria, viruses, fungi, and other microbes are part of a growing list of pathogens found in the brains of patients with neurodegenerative diseases (NDDs), including Lyme *neuroborreliosis*.

Microbes in the brain may indicate meningitis or encephalitis, two diseases that are active infections with inflammation. For diseases like Parkinson's (PD), Alzheimer's (AD) and other NDDs that were not thought to be infectious, finding pathogens in the brain is both surprising and concerning. Table 14.2 lists the various pathogens found so far in the brain.

Table 14.2 – Pathogens in the brain

Pathogen	Origin/Cause	Effect
***Porphyromonas gingivalis (P. gingivalis)* bacterium**	Mouth	Some of the proteins made by this bacterium have been found in the brain
***Fusobacterium nucleatum* bacterium**		
***Prevotella intermedia* bacterium**		
Herpes simplex virus (HSV)	Sexually transmitted	o Lives for years in nerve cells that supply the face and lips. o Can migrate back up the same route into the brain producing mild inflammatory response
HSV-1, 2	Shed by an infected person or sexually transmitted	o Shows higher levels in Alzheimer patients o Can lead to multiple sclerosis o Can cause encephalitis, meningitis, and disseminated infection in immuno-compromised patients o Possible link to neurodegenerative diseases (Parkinson's, Alzheimer's)
Measles virus	o Exposure to an infected person (cough, sneezes)	o Can cause multiple sclerosis and subacute sclerosing panencephalitis

Lyme disease

	o Contact through saliva or nasal secretions	(SSP)
HIV-1, 2	Sexual contact with infected person	Can cause dementia, Parkinson-like and Alzheimer-like disease, and encephalitis
H1N1 virus (HSV-1, 2)	Possibly originating in the gut, manifesting as digestive tissues, and then moving indirectly into the brain as it cannot penetrate the blood-brain barrier (BBB)	o Delayed Parkinson's disease (controversial) o *Encephalitis lethargica* (EL) – a possible precursor of Alzheimer's disease o May not cause Parkinson's disease, but may prime the central nervous system and, with the addition of toxin(s), lead to Parkinson's disease o May cause central central nervous system immune cells (the microglia) to flow into the *substantia nigra* and the hippocampus, causing inflammation and cell death in the area
H5N1 virus (a subset of H1N1)	Possibly originating in the gut, manifesting as digestive tissues, and then moving indirectly into the brain by infecting neurons first in the gut, then into the vagus nerve, and subsequently into the *substantia nigra*	o Parkinsonism (symptoms: brain inflammation, tremors, other motor functions) o May degenerate into Parkinson's disease
Fungi *aspergilii*	Unclear	Brain infection as cysts
Protozoa toxoplasma gondii		
Parasites (Taenia solium, pork tapeworm)		
Syphilis	o *Triponema pallidum* (a spirochete type of bacterium o Sexual transmission o Syphilis is transmitted from mother to baby during pregnancy or at birth	Can live in the brain for decades, eventually infecting dementia and causing dementia
Lyme disease	o *Borrelia burgdorferi* carried by the deer tick *Ixodes scapularis* o Bacterium also spread by rodents	o Infectious disease causing a rash that expands from the site of infection on the skin o Causes fever, headache, and tiredness. If untreated, may lead to the loss of ability to move one or both sides of the face, joints pain, severe headaches with neck stiffness or heart palpitations, memory problems
Ehrlichia	Not clear	Infects white blood cells
***Babesia* (relative of the malaria parasite)**	Not clear	Infects red blood cells
Bartonella	Not clear	Infects blood vessels

	Lyme disease	
Alzheimer's disease	Many different organisms. Also by sterile inflammation, not from invading pathogens	Also harbors fungi

Source: Fymat (2018, 2019)

How do the organisms in Table 14.2, and others, get into the brain since it is protected by the blood-brain barrier? They do so when the barrier looses some of its impermeability. Other avenues for reaching directly the brain are: (a) intra-nasal and sinus access, (b) the mouth (through the lingual nerve, which runs down the jawline and into the tongue), (c) the gut (through the vagus nerve, which travels through the neck and thorax to the stomach), and even (d) the eyes (through the olfactory bulbs), all of which connect to the brain by replicating and spreading.

As seen (at least in part from Table 14.2), brain infections can be caused by viruses, bacteria, fungi or, occasionally, protozoa or parasites. (Another group of brain disorders, not discussed here, are the spongiform encephalopathies, which are caused by abnormal proteins called prions.) Infections of the brain often also involve other parts of the central nervous system (CNS), including the spinal cord. They can cause inflammation of the brain (encephalitis), and the layers of tissue (meninges) that cover the brain and the spinal cord (meningitis). Often, bacterial meningitis spreads to the brain itself, causing encephalitis. Similarly, viral infections that cause encephalitis often also cause meningitis. Usually, in encephalitis and meningitis, infection is not confined to one area but may occur throughout the brain, within the meninges, along the entire length of the spinal cord, and over the entire brain. Infection may also be confined to one area (empyema in an existing space in the body; abscess).

Sometimes, a brain infection, a vaccine, cancer, or another disorder may trigger a misguided immune reaction, causing the immune system to attack normal cells in the brain (an autoimmune reaction) as a result of which the brain becomes inflamed (a so-called post-infectious encephalitis).

In sum, brain infections may manifest as a diffuse infection resulting in encephalitis, sometimes affecting specific areas on the brain, according to two processes: (a) inflammation of the brain secondary to meningeal or parameningeal infections or (b) focal infection (e.g., due to a brain abscess or to fungal or parasitic brain infections). Encephalitis is most commonly due to viruses (herpes simplex, herpes zoster, cytomegalovirus, West Nile virus, HIV, and prion disease).

Bacteria and other infectious organisms can reach the brain and meninges in several ways by being carried by the blood, entering the brain directly from the outside (for example, through a skull fracture or during surgery on the brain), or spreading from nearby infected structures (for example, sinuses or middle ear). Some viruses can enter the body through the nose and mouth and move to the brain by replicating and spreading through the olfactory bulbs, the lingual nerve or the vagus nerve, as indicated earlier. Viral infections may cause brain damage but we are uncertain whether the injuries play a role in NDDs although some studies do suggest a connection.

When interacting with the nervous system, viral particles can cross the BBB in a number of ways:

- **Direct crossing:** through infection of endothelial cells;

- **Trojan horse approach:** by infecting monocytes that cross the BBB before replicating and bursting out of the white blood cells once inside the brain;

- **Invoking an immune response:** Not crossing the BBB directly but invoking an immune response that may spur cytokines or chemokines to breach the BBB.

Once inside the brain, viruses can infect cells or their myelin sheath and kill them. They do not necessarily have to enter the brain to cause damage, though. They can also spark an immune response that activates microglia which then consume otherwise healthy neurons.

The connections between measles and herpes viruses with multiple sclerosis (MS), and between measles, herpes, and HIV viruses with Alzheimer's-like and Parkinson's-like disease are explored in detail in my other publications (see Fymat 2018, 2019, 2020).

In summary, viral (and other pathogen) infections can damage the brain but there is currently no definitive study demonstrating that a virus (or other pathogen) can cause PD or AD or any number of other NDDs. Viruses and other pathogens can cause a lot of different brain diseases... but this remains only a hypothesis whose rigorous testing is long overdue. A major roadblock in definitively establishing the link between pathogens and NDDs remains our lack of understanding of the long-delayed onset of the disease after the infection/inflammation. Further, even if a definitive link were established, it may only be correlative not causal. At this juncture, I assert that the root cause of AD (and other NDDs) may be a runaway autoimmune disease that is unable to maintain homeostasis between opposing synaptoblastic and synaptoclastic pressures, everything else being the consequences of risks, correlations,...remains the only plausible explanation.

The following sidebars, Sidebars 14.1 and 14.2, provide an overview of Herpes simplex viruses that infect the brain and familial and autoimmune encephalitis, respectively.

Take-home points

- Cytomegalovirus is one of the co-infections of Lyme disease that are most frequently cited, however, it is but one of several brain pathogens. It is a member of the Herpes family. It shares with it the common ability to remain alive, yet dormant, in the human body for the life of its human host.

- Approximately 80% of adults in the U.S. and most people across the globe are infected with cytomegalovirus.

- Rarely does cytomegalovirus become active unless the immune system is weakened and rendered unable to hold the virus in check.

- Bacteria, viruses, fungi, and other microbes are part of a growing list of pathogens found in the brains of patients with neurodegenerative diseases, including Lyme *neuroborreliosis.*

Lyme disease

- Microbes in the brain may indicate meningitis or encephalitis, two diseases that are active infections with inflammation. For diseases like Parkinson's (PD), Alzheimer's (AD) and other neurodegenerative diseases that were not thought to be infectious, finding pathogens in the brain is both surprising and concerning.

- Brain pathogens of interest are those for Lyme, *Ehrlichia*, *Babesia,* and *Bartonella*. These and other organisms get into the brain when the blood-brain barrier looses some of its impermeability or by directly reaching the brain through intra-nasal and sinus access, the mouth (through the lingual nerve), the gut (through the vagus nerve), and even the eyes (through the olfactory bulbs).

- Brain infections may manifest as a diffuse infection resulting in encephalitis, sometimes affecting specific areas on the brain following focal infection and inflammation of the brain.

- Viral (and other pathogen) infections can damage the brain but there is currently no definitive study demonstrating that a virus (or other pathogen) can cause Parkinson's disease or Alzheimer's disease or a number of other neurodegenerative diseases.

Sidebar 14.1

Overview of Herpes simplex viruses that infect the brain

Table 14.3 lists the eight types of herpes simplex viruses that infect humans (HHSV 1-8). Of these, types 1, 2, 4 and 6 might play a role in NDDs. After initial infection, they all remain latent but may subsequently reactivate and transmit to other people. Because they do not survive long outside a host, their transmission usually requires intimate contact with a person shedding the virus. Viral shedding occurs from lesions but can occur even when lesions are not apparent. This very contagious viral infection is spread by direct contact with sores or sometimes contact with an affected area when no sores are present. After the initial infection, HSV remains dormant in nerve ganglia, from which it can periodically emerge, causing symptoms.

Epstein-Barre virus (EBV), HHV 4 and HHV 8 - also known as Kaposi sarcoma–associated herpes virus (KSHV), can cause certain cancers, a topic not considered here.

Table 14.3 - Eight types of Herpes viruses infect humans

Common name	Other name	Manifestation(s)
Herpes simplex virus-mimicking type 1	Human herpes virus type 1	Gingivitis, keratoconjunctivitis, cutaneous herpes, genital herpes, herpes labials, encephalitis, viral meningitis, esophagitis, pneumonia, disseminated

Lyme disease

		infection, hepatitis
Herpes simplex virus-mimicking type 2	Human herpes virus type 2	Genital herpes, cutaneous herpes, neonatal herpes, viral meningitis, disseminated infection, hepatitis
Varicella zoster virus	Human herpes virus type 3	Chickenpox, herpes zoster, disseminated herpes zoster
Epstein-Barr virus	Human herpes virus type 4	Infectious mononucleosis, hepatitis, encephalitis, nasopharyngeal carcinoma, Hodgkin's lymphoma,. Burrito's lymphoma, non-proliferation syndromes, oral hairy leukoplakia
Cytomegalovirus	Human herpes virus type 5	CMV mononucleosis, hepatitis, congenital cytomegalic inclusion disease, retinitis, pneumonia, colitis
Human herpes virus type 6	Human herpes virus type 6	Roseola infant, otitis media with fever, encephalitis
Human herpes virus type 7	Human herpes virus type 7	Roseola infant
Kaposi sarcoma associated herpes virus	Human herpes virus type 8	Not a known cause of acute illness but has a causative role in Kaposi sarcoma, and AIDS-related non-Hodgkin lymphoma that grow primarily in the plural, epicardial, or abdominal cavities as lymphomatous effusions. Also linked with multicultural Cattleman's disease

Source: Kenneth M. Kaye, Merck Manual

Sidebar 14.2
Familial and autoimmune encephalitis

Encephalitis is inflammation of the brain that occurs when a virus, vaccine, or something else triggers inflammation. The spinal cord may also be involved, resulting in a disorder called encephalomyelitis. Because inflammation plays a role in the development of NDDs, the consideration of encephalitis is appropriate. It is most commonly due to viruses, such as HSV 1, 2, varicella zoster, cytomegalovirus, or West Nile virus. It can occur in the following ways: (a) a virus directly infects the brain; (b) a virus that caused an infection in the past becomes reactivated and directly damages the brain; and (c) a virus or vaccine triggers a reaction that makes the immune system attack brain tissue (an autoimmune reaction).

Sometimes, bacteria cause encephalitis, usually as part of bacterial meningitis (called meninges-encephalitis). Brain infections due to an autoimmune reaction sometimes develop in people who have cancer. Protozoa—such as amoebas, the protozoa that cause toxoplasmosis (in people who have AIDS), and those that cause malaria can also infect the brain and cause encephalitis.

Interestingly, infections that can directly lead to encephalitis can occur in epidemics as, for example, the 1918 Spanish H1N1 flu epidemic.

Types of encephalitis

Encephalitis spread by viruses

There are six types of viral encephalitis, including:

- *La Crosse encephalitis:* It is caused by the La Crosse virus (also called California virus). It accounts for most cases in children. Many cases are mild and undiagnosed. Fewer than 1% of infected people die from it.

- *Eastern equine encephalitis:* It affects mainly young children and people older than 55. In children younger than 1 year, it can cause severe symptoms and permanent nerve or brain damage. Over half of infected people die.

- *West Nile encephalitis:* This virus also causes a milder infection called West Nile fever, which is much more common. West Nile encephalitis develops in fewer than 1% of people who develop West Nile fever. About 9% of people with West Nile encephalitis die. However, those who have only West Nile fever usually recover fully.

- *St. Louis encephalitis:* Infection is more likely to affect the brain in older people. Epidemics once occurred about every 10 years but are now rare.

- *Western equine encephalitis:* For unknown reasons, this type of encephalitis has largely disappeared since 1988. It can affect all age groups but is more severe and more likely to affect the brain in children younger than 1 year.

- *Sassanian virus encephalitis:* It usually causes mild or no symptoms and has ben associated with cases of encephalitis. The virus is similar to the one that causes tick-borne encephalitis. However, the infection can also cause severe encephalitis with headache, vomiting, seizures, loss of coordination, speech problems, or coma. Up to 15% of people with severe encephalitis die. The vaccine that is effective against tick-borne encephalitis is not effective against the Sassanian virus.

- **Encephalitis spread by ticks:** These two types of encephalitis are of more direct interest here. They include:

  - *Tick-borne encephalitis:* It usually causes a mild flu-like illness that clears up within a few days. A vaccine is available.

  - *Colorado tick fever:* It is a flu-like illness. Occasionally, it causes meningitis or encephalitis.

- **Encephalitis spread by mosquitoes:** Several viruses that cause encephalitis were once present in only a few parts of the world but now are spreading, probably because travel has increased. These viruses include the:

    - *Chikungunya virus:* It can lead to severe encephalitis and even death, especially in infants and people over age 65.

    - *Japanese encephalitis virus:* A common cause in Asia and the Western Pacific.

    - *Venezuelan equine encephalitis:* It occurs mainly in travelers returning from areas where the virus is common.

    - *Zika virus:* It may cause fever, joint and muscle aches, headache, and a red, bumpy rash. Having Zika virus infection during pregnancy can cause microcephaly and severe brain damage in the baby.

Autoimmune encephalitis

Encephalitis can result from reactivation of a virus, including:

- *HSV 1.*

- *HHV 3.*

- *JC virus:* It causes a usually fatal disorder called progressive multifocal leukoencephalopathy (PML) that is common among people who have AIDS or other conditions that impair the immune system; and

- *Measles virus:* If reactivated, it leads to a usually fatal disorder called subacute sclerosing panencephalitis (SSP) years after measles occurs.

A reactivated infection, which can occur long after people had the infection, can severely damage the brain.

15

Epidemiology of Lyme

The following data have been extracted from publications by the (U.S.) Centers for Disease Control & Prevention (CDC&P). Possible cases of LD are reported to State and local Health Departments by health care providers and laboratories. In turn, State Health Departments classify cases according to standard criteria outlined in the LD case definition, and report "confirmed" and "probable" cases to CDC&P. The extent of case investigations varies by State. Investigations are often dependent on available resources and staff time. Some States describe their surveillance methods in detail on their Health Department website. It is always a good idea to consult such websites whenever traveling to areas where LD possibly occurs.

Figure 15.1 highlights the geographical areas within the U.S. that are infected by LD. The densities of the most vulnerable areas are evidenced by the darker purple colors. In the less affected areas, the purple dots are scattered and isolated.

Each year, approximately 30,000 cases of LD are reported to CDC&P by State Health Departments and the District of Columbia. However, this number does not reflect every case of LD that is diagnosed in the country. Standard national surveillance is only one way that public health officials can track where a disease is occurring and with what frequency. Only a limited selection of data is presented below; more data can be found in the website of CDC&P.

Recent Surveillance Data

In 2017, a total of 42,743 confirmed and probable cases of LD were reported to CDC&P, over 17% more than in 2016. The geographic distribution of high incidence areas with LD appears to be expanding based on data reported to the National Notifiable Disease Surveillance System (NNDSS). The number of counties with an incidence of ≥ 10 confirmed cases per 100,000 persons increased from 324 in 2008 to 454 in 2017. The annual reported cases of LD are shown in Figure 15.2 for the time period (1991-2017).

Lyme disease

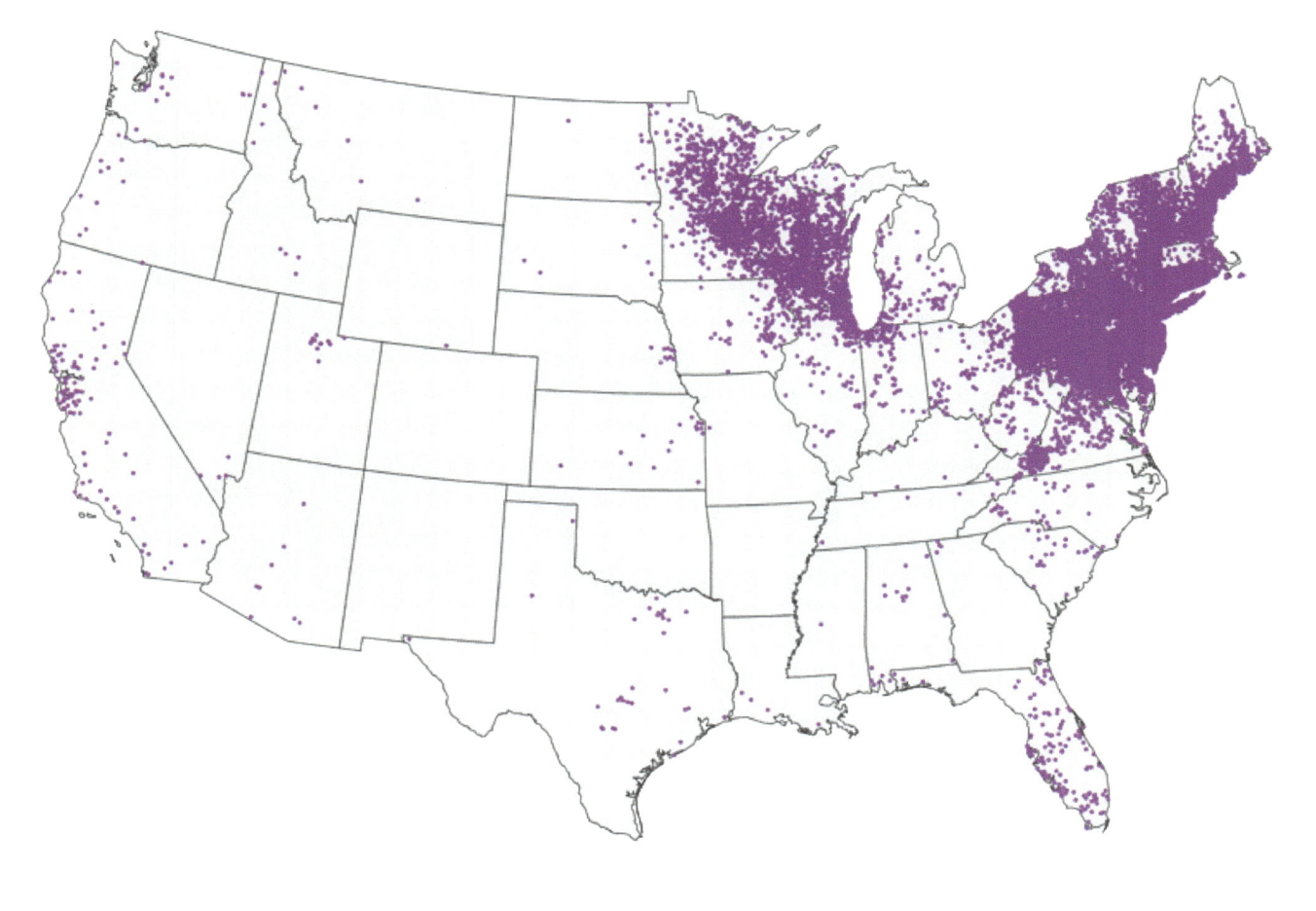

Figure 15.1 – Geographical areas of the U.S. affected by Lyme disease

Source: CDC&P

Figure 15.2 below graphs the annual reported cases of Lyme disease for 1991-2017. While that number varies from year-to-year, it has essentially been climbing during that 27-year time period, quadrupling from approximately 10,000 to 40,000. If that trend continues, by year 2044, the number of cases will climb to 160,000.

On the other hand, Table 15.1 shows the LD surveillance data by State. For each State, it lists the incidence rate, the numbers of incident, confirmed, and probable cases for year 2017, and the 3-year incidence. It is readily seen from the Table that the States of high incidence rate are: Connecticut, Delaware, Maine, Maryland, Massachusetts, Minnesota, New Hampshire, New Jersey, New York, Pennsylvania, Rhode Island, Vermont, Virginia, Washington D.C., West Virginia, and Wisconsin, all States lying in the north eastern region of the country.

Lyme disease

Figure 15.2 – Annual reported cases of Lyme disease (1991-2017)

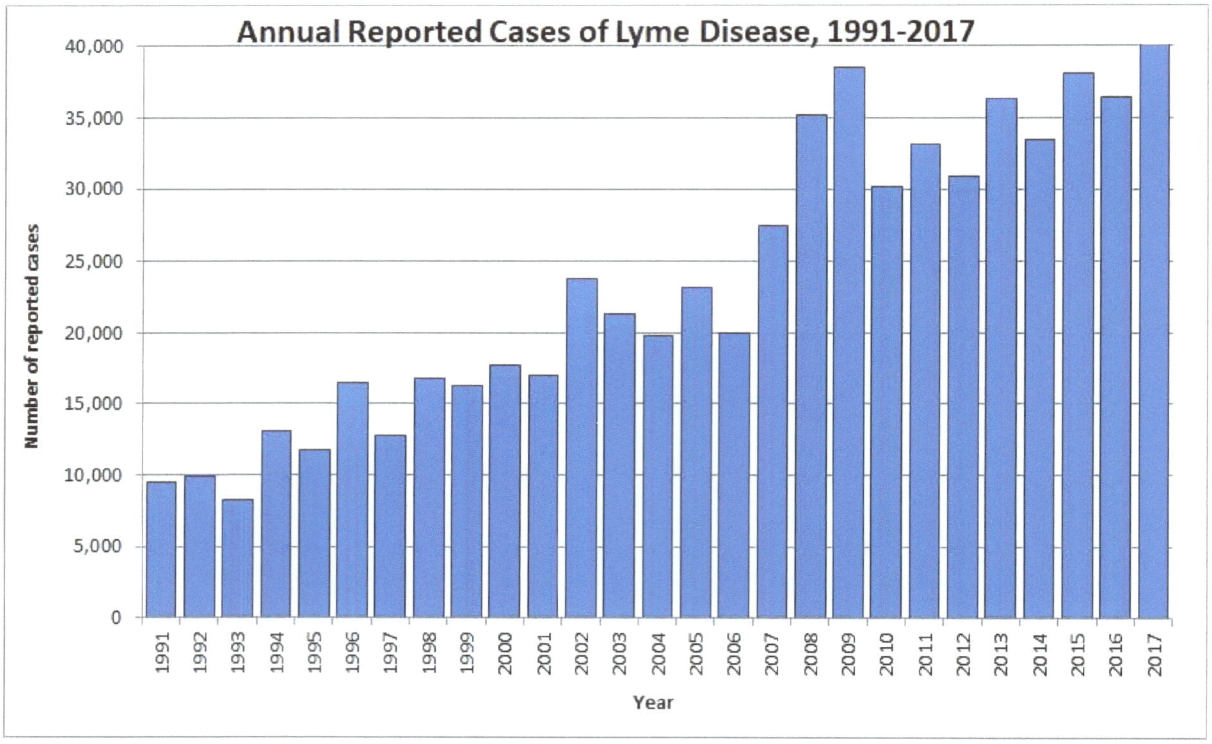

Source: CDC&P

Table 15.1 – Lyme disease surveillance data by State

Location	Incidence rate	2017 confirmed	2017 probable	2017 incidence	3-year incidence
Alabama	Low	25	16	0.5	0.4
Alaska	Low	8	2	1.1	0.7
Arizona	Low	18	10	0.3	0.2
Arkansas	Low	2	4	0.1	0
California	Low	84	61	0.2	0.2
Colorado	Low	4	0	0.1	0
Connecticut	High	1381	670	38.5	41.8
Delaware	High	421	187	43.8	40
Florida	Low	124	86	0.6	0.6
Georgia	Low	8	0	0.1	0.1

Lyme disease

State	Level				
Hawaii	Low	0	0	0	0
Idaho	Low	15	5	0.9	0.5
Illinois	Low	218	55	1.7	1.9
Indiana	Low	92	50	1.4	1.6
Iowa	Low	112	143	3.6	3.4
Kansas	Low	13	27	0.4	0.5
Kentucky	Low	9	14	0.1	0.3
Louisiana	Low	8	4	0.2	0.1
Maine	High	1424	426	106.6	89.2
Maryland	High	1194	697	19.7	26
Massachusetts	High	321	89	4.7	16.6
Michigan	Low	196	95	2	1.6
Minnesota	High	1408	1910	25.2	23.4
Mississippi	Low	1	0	0	1
Missouri	Low	2	10	0	0
Montana	Low	4	8	0.4	0.6
Nebraska	Low	6	8	0.3	0.4
Nevada	Low	7	10	0.2	0.2
New Hampshire	High	956	425	71.2	51.9
New Jersey	High	3629	146.3	40.3	40.5
New Mexico	Low	3	0	0.1	0
New York	High	3502	1653	17.6	15.8
North Carolina	Low	70	22.5	0.7	0.5
North Dakota	Low	12	44	1.6	1.6
Ohio	Low	233	37	2	1.3
Oklahoma	Low	0	1	0	0
Oregon	Low	13	71	0.3	0.3
Pennsylvania	High	9250	2650	72.2	68.7
Rhode Island	High	595	537	56.2	51.4
South Carolina	Low	1.3	8	0.3	0.3

Lyme disease

South Dakota	Low	4	8	0.5	0.7
Tennessee	Low	15	32	0.2	0.1
Texas	Low	31	22	0.1	0.1
Utah	Low	10	15	0.3	0.3
Vermont	High	646	446	103.6	86.7
Virginia	High	1041	616	12.3	12.3
Washington	Low	26	11	0.4	0.3
Washington D.C.	High	62	22	8.9	10.1
West Virginia	High	503	145	27.7	19
Wisconsin	High	1794	1206	31	26.6
Wyoming	Low	3	1	0.5	0.2

Source: Data from the (U.S.) Centers for Disease Control & Prevention (CDC&P), National Center for Emerging and Zoonotic Infectious Diseases (NCEZID), Division of Vector-Borne Diseases

LD by age groups

The reported cases in 2017 by age groups are summarized in Table 15.2. Data are provided for both confirmed and probable cases. For confirmed cases, the peaks occur at ages 5-9 and 60-64 with a trough at ages 20-24. For probable cases, the peaks occur at the same age groups as for the confirmed cases but the trough occurs at the different age group 40-44. The difference could be explained if it is recalled that the probable cases are the result of probability considerations, which may not (and actually do not) accurately reflect reality. For both confirmed and probable cases, the lowest values occur at ages 85+.

Table 15.2 – Reported cases of LD by age groups (2017)

Age group	Confirmed cases	Probable cases
0-4	1,330	274
5-9	2,686	850
10-14	2,130	823
15-19	1,420	678
20-24	1,080	623
25-29	1,104	622
30-34	1,222	618
35-39	1,398	606

Lyme disease

Age group		
40-44	1,296	590
45-49	1,705	742
50-54	2,227	892
55-59	2,596	1,121
60-64	**2,639**	**1,197**
65-69	2,324	1,155
70-74	1,691	903
75-79	1,109	657
80-84	548	376
85+	**362**	**271**

The above results are more conveniently graphed in Figures 15.1 and 15.2.

LD by gender and age groups

The data by gender for the same age groups are shown in Table 15.3 among reported cases (confirmed and probable). The disease incidence is generally larger for men than women for both probable and confirmed cases. It is both maximal and minimal at the same age group 5-9 in all cases. With advancing age, it is maximal for both genders for confirmed cases at age group 65-69 whereas, for probable cases, it occurs at the later age group 75-79. At age 85+, there is a surprising secondary maximum for males but not for females. Further, for females, the confirmed cases are lowest at the age group 85+ whereas, for males, there is also a minimum for that same age group but it is not the lowest.

Table 15.3 – LD rates among reported (confirmed and probable) by age group and sex (2017)

Age group	Male confirmed incidence	Male probable incidence	Female confirmed incidence	Female probable incidence
0-4	6.83	1.44	6.35	1.29
5-9	**15.72**	**4.80**	**10.39**	**3.32**
10-14	13.29	4.54	6.80	3.25
15-19	8.52	3.40	4.57	2.90
20-24	5.89	2.95	3.64	2.55
25-29	**5.69**	**2.81**	**3.57**	**2.39**
30-34	6.31	3.08	4.59	2.42
35-39	7.83	3.17	5.07	2.46
40-44	7.76	3.49	5.27	2.42

Lyme disease

45-49	9.60	3.83	6.49	3.06
50-54	11.37	4.67	9.16	3.50
55-59	13.42	6.08	9.89	4.08
60-64	15.19	7.52	11.01	4.44
65-69	**15.98**	8.49	**11.45**	5.32
70-74	15.02	8.79	11.13	5.43
75-79	14.29	**9.34**	11.03	**5.88**
80-84	11.40	8.21	7.35	4.86
85+	7.19	**9.79**	3.4	3.27

Take-home points

- Possible cases of Lyme disease are reported to State and local Health Departments by health care providers and laboratories. In turn, State Health Departments classify cases according to standard criteria and report "confirmed" and "probable" cases to the (U.S.) Center for Disease Control & Prevention.

- Some States describe their surveillance methods in detail on their Health Department website. It is always a good idea to consult such websites whenever traveling to areas where Lyme disease possibly occurs.

- Each year, approximately 30,000 cases of Lyme disease are reported. However, this number does not reflect every case that is diagnosed.

- In 2017, a total of 42,743 confirmed and probable cases of Lyme disease were reported, which is over 17% more than in 2016.

- The geographic distribution of high incidence areas with Lyme disease appears to be expanding with the number of counties with an incidence of ≥10 confirmed cases per 100,000 persons having increased from 324 in 2008 to 454 in 2017.

- The States of high incidence rate are: Connecticut, Delaware, Maine, Maryland, Massachusetts, Minnesota, New Hampshire, New Jersey, New York, Pennsylvania, Rhode Island, Vermont, Virginia, Washington D.C., West Virginia, and Wisconsin, all States lying in the north eastern region of the country.

- By age groups, the reported confirmed and probable cases in 2017 showed peaks occurring at age groups 5-9 and 60-64. The lowest values occur at ages 85+.

- The data by gender and age groups among reported cases (confirmed and probable) show a disease incidence that is generally larger for men than women. However, with some nominal departures, the variation with age for both genders is essentially the same. Peak and minimum incidence occur essentially at the same age groups for both genders.

16

Frequently asked questions

The following series of frequently asked questions has been prepared by the (U.S.) Center for Disease Control & Prevention (CDC&P). It has been subdivided into the following categories of questions: general; transmission; diagnosis, testing, and treatment; surveillance of the disease; and CDC&P specific questions. The following has been largely borrowed from this material.

General questions

1. Can I get LD anywhere in the U.S.?

No. LD is spread through the bite of a black-legged tick (named *Ixodes scapularis* or *Ixodes pacificus*) that is infected with *Borrelia burgdorferi*. In the U.S., most infections occur in the following endemic areas:

- Northeast and mid-Atlantic States, from northeastern Virginia to Maine;
- North central States, mostly in Wisconsin and Minnesota; and
- West Coast, particularly northern California.

Maps showing the distribution of human cases are based on where people live. Because people travel, these travel destinations are not necessarily where people became infected. Cases are sometimes diagnosed and reported from an area where LD is not expected, but they are almost always travel-related.

2. Having been bitten by a tick, do I have LD?

The chance that you might get LD from a single tick bite depends on the type of tick, where you acquired it, and how long it was attached to you. Many types of ticks bite people in the U.S., but only black-legged ticks transmit the bacteria that cause LD. Furthermore, only these ticks in the highly endemic areas of the northeastern and north central U.S. are commonly infected.

Black-legged ticks need to be attached for at least 24 hours before they can transmit LD. This is why it is so important to remove them promptly (with fine-tipped tweezers) and to check your body daily for ticks if you live in an endemic area. If you develop illness within a few weeks of a tick bite, see your health care provider right away. Common symptoms of LD include: a rash, fever, body aches, facial paralysis, and arthritis. Ticks can also transmit other diseases, so it is important to be alert for any illness that follows a tick bite.

3. If I have LD, is it possible that I also have other tick-borne diseases?

Black-legged ticks can spread germs that cause LD and several other tick-borne diseases. A person who has more than one tick-borne disease at a time is said to have a co-infection. The frequency of co-infections varies widely from place-to-place and over time.

The most common co-infections that occur with LD are *anaplasmosis* and *babesiosis*. In general:

- Co-infection with LD and *anaplasmosis* happens from 2% to 12% of the time;

- Other co-infections include *babesiosis,* Powassan virus disease, and *Borrelia miyamotoi* disease. They occur less frequently. Additional research is needed to know how often these co-infections occur.

LD and *anaplasmosis* are treated with the same antibiotic, so a person getting treatment for LD will be treated for *anaplasmosis* at the same time, regardless of whether additional tests were run.

Babesiosis is a parasitic disease that is treated with different medications. If your LD symptoms do not seem to be going away after taking antibiotics, see your healthcare provider. Although some providers also test patients for *Bartonella* or mycoplasma co-infections, there is no evidence that these germs are spread by ticks.

If you have been diagnosed with co-infections, you may consider getting a second opinion. CDC&P recommends finding a board-certified infectious disease specialist, internist, or pediatrician affiliated with a university teaching hospital. Additionally, your State Health Department is typically the best source of information about tick-borne diseases that occur in your area.

4. In the Southern U.S. with a lot of lone star tick bites, can I get Southern Lyme disease?

The lone star ticks (*Amblyomma americanum*) are primarily found in the southeastern and eastern U.S. They do not transmit LD. However, you are correct to be concerned about this very aggressive species as it can spread *ehrlichiosis*, *tularemia*, and Southern tick-associated rash illness (STARI). The STARI rash is a red, expanding "bull's eye" lesion that develops around the site of a lone star tick bite. The rash usually appears within seven days of a tick bite and expands to a diameter of 8 centimeters (3 inches) or more. The rash should not be confused with much smaller areas of redness and discomfort that can occur commonly at tick bite sites. Unlike LD, STARI has not been linked to arthritis, neurological problems, or chronic symptoms.

Nevertheless, the similarity between the STARI bull's eye rash and the LD bull's eye rash has created much public confusion. The pathogen responsible for STARI has not been identified.

In contrast, LD in North America is caused by a specific type of bacteria, *Borrelia burgdorferi*, which is transmitted by two species of black-legged ticks, *Ixodes scapularis* and *Ixodes pacificus*. While black-legged ticks exist in the southern U.S., their feeding habits in this region make them much less likely to maintain, sustain, and transmit LD.

5. Is it true that if I have LD, I will always have it?

No. Patients treated with antibiotics in the early stages of the infection usually recover rapidly and completely. Most patients who are treated in later stages of the disease also respond well to antibiotics, although some may have suffered long-term damage to the nervous system or joints. It is not uncommon for patients treated for LD with a recommended 2-4 week course of antibiotics to have lingering symptoms of fatigue, pain, or joint and muscle aches at the time they finish treatment. In a small percentage of cases, these symptoms can last for more than 6 months. These symptoms cannot be cured by longer courses of antibiotics, but they generally improve on their own, over time.

6. I am pregnant and think I might have LD, what should I do?

If you are pregnant and suspect you have contracted LD, contact your physician immediately. Untreated LD during pregnancy can lead to infection of the placenta. Although rare, spread from mother to fetus is possible. Fortunately, with appropriate antibiotic treatment, there is no increased risk of adverse birth outcomes. There are no published studies assessing developmental outcomes of children whose mothers acquired Lyme disease during pregnancy.

7. What is CLD?

LD is caused by infection with the bacterium *Borrelia burgdorferi*. Although most cases of LD can be cured with a 2-4 week course of oral antibiotics, patients can sometimes have symptoms of pain, fatigue, or difficulty thinking that last for more than 6 months after they finish treatment. This condition is called post-treatment Lyme disease syndrome (PLDS). The term "chronic Lyme disease" (CLD) is also sometimes used; however, this term has been used to describe a wide variety of different conditions and, therefore, can be confusing. Because of the confusion in how the term CLD is employed, experts do not support its use.

Transmission questions

1. Can LD be transmitted during a blood transfusion?

Although no cases of LD have been linked to blood transfusion, scientists have found that the LD bacteria can live in blood from a person with an active infection that is stored for donation. Individuals being treated for LD with an antibiotic should not donate blood. Individuals who have completed antibiotic treatment for LD may be considered as potential blood donors. The Red Cross provides additional information on the most recent criteria for blood donation.

2. Can LD be transmitted sexually?

There is no credible scientific evidence that LD is spread through sexual contact. Published studies in animals do not support sexual transmission, and the biology of the LD spirochete is not compatible with this route of exposure.

The ticks that transmit LD are very small and easily overlooked. Consequently, it is possible for sexual partners living in the same household to both become infected through tick bites, even if one or both partners does not remember being bitten.

3. Can LD be transmitted through breast milk?

There are no reports of LD being spread to infants through breast milk. If you are diagnosed with LD and are also breastfeeding, make sure that your doctor knows this so that he/she can prescribe an antibiotic that is safe for use when breast-feeding. However, LD can be transplanted from the mother to the placenta. Although known and documented for approximately the last 35 years especially in Canada, this fact has only been recognized in the U.S. very recently in 2019.

Diagnosis, testing, and treatment questions

1. Are the diagnostic tests recommended by CDC&P accurate? If not, can I be treated based on my symptoms or do I need different tests?

You may have heard that the blood test for LD is correctly positive only 65% of the time or less. This is misleading information. As with serologic tests for other infectious diseases, the accuracy of the test depends upon how long you have been infected. During the first few weeks of infection, such as when a patient has an *erythema migrans* rash, the test is expected to be negative. Several weeks after infection, FDA-cleared tests have very good sensitivity.

2. Is it possible for someone who was infected with LD to test negative?

Yes, because:

- Some people who receive antibiotics (e.g., doxycycline) early in disease (within the first few weeks after tick bite) may not have a fully developed antibody response or may only develop an antibody response at levels too low to be detected by the test; and

- Antibodies against LD bacteria usually take a few weeks to develop, so tests performed before this time may be negative even if the person is infected. In this case, if the person is retested a few weeks later, they should have a positive test if they have LD. It is not until 4-6 weeks have passed that the test is likely to be positive. This does not mean that the test is bad, only that it needs to be used correctly.

3. I have been sick for a few years with joint and muscle pain, fatigue, and difficulty thinking. I was tested for LD using a Western blot test. The IgM test was positive but the IgG test was negative. Is LD the cause of my symptoms?

Probably not. First, you should only have an immunoblot (such as an FDA-approved Western Blot or striped blot) test done if your blood has already been tested and found reactive with an enzyme immunoassay (EIA) or immunofluorescent assay (IFA). Second, the IgM Western Blot test result is only meaningful during the first four weeks of illness. If you have been infected for longer than 4-6 weeks and the IgG Western Blot is still negative, it is highly likely that the IgM result is incorrect (e.g., a false positive). This does not mean that you are not ill, but it does suggest that the cause of illness is something other than the LD bacterium.

4. Where can I get a test to make sure I am cured?

As with many infectious diseases, there is no test that can "prove" cure. Tests for LD detect antibodies produced by the human immune system to fight off the bacteria (*Borrelia burgdorferi*) that cause LD. These antibodies can persist long after the infection is gone. This means that if your blood tests positive, then it will likely continue to test positive for months or even years even though the bacteria are no longer present.

A research tool called polymerase chain reaction (PCR) can detect bacterial DNA in some patients. Unfortunately, this is also not helpful as a test of whether the antibiotics have killed all the bacteria. Studies have shown that DNA fragments from dead bacteria can be detected for many months after treatment. Studies have also shown that the remaining DNA fragments are not infectious. Positive PCR test results are analogous to a crime scene – just because a robbery occurred and the robber left his DNA, it does not mean that the robber is still in the house. Similarly, just because DNA fragments from an infection remain, it does not mean the bacteria are alive or viable.

5. Even though I finished three weeks of antibiotic treatment, my blood test for LD is still positive. Am I still infected?

No. The tests for LD detect antibodies made by the immune system to fight off the bacteria, *Borrelia burgdorferi*. Your immune system continues to make the antibodies for months or years after the infection is gone. This means that once your blood tests positive, it will continue to test positive for months to years even though the bacteria are no longer present. Unfortunately, in the case of bacterial infections, these antibodies do not prevent someone from getting LD again if they are bitten by another infected tick.

6. Can you recommend a doctor who is familiar with diagnosing and treating LD?

In areas where LD is common, most family practice physicians, general practitioners, and pediatricians are familiar with diagnosing and treating it. If you have symptoms that suggest LD, or any other tick-borne infection, tell your doctor all these facts. Many doctors may not consider tick-borne diseases in diagnosing your illness unless you report being bitten by a tick, or live in, or have recently visited, a tick-infested area.

In areas where LD is not common or for more complicated cases of LD, infectious disease specialists are often the best type of doctor to see. Please note that CDC&P cannot evaluate the qualifications and competence of individual doctors; however, the (U.S.) National Institutes of Health provides information about how to choose a doctor. Additionally, your State Medical Board can help you find out if your health provider is in good standing.

Surveillance questions

1. How many people get LD?

Each year, approximately 30,000 cases of LD are reported to CDC&P by state health departments and the District of Columbia. However, this number does not reflect every case of LD that is diagnosed in the U.S. every year.

Surveillance systems provide vital information but they do not capture every illness. Because only a fraction of illnesses are reported, researchers need to estimate the total burden of illness to set public health goals, allocate resources, and measure the economic impact of disease. CDC&P uses the best data available and makes reasonable adjustments—based on related data, previous study results, and common assumptions—to account for missing pieces of information.

To improve public health, CDC&P wants to know how many people are actually diagnosed with LD each year and for this reason has conducted two studies:

- Study # 1: ("Lyme Disease Testing by Large Commercial Laboratories in the United States"): The study estimated the number of people who tested positive for LD based on data obtained from a survey of clinical laboratories. Researchers estimated that 288,000 (range 240,000–444,000) infections occur among patients for whom a laboratory specimen was submitted in 2008.

- Study # 2: ("Incidence of Clinician-Diagnosed Lyme Disease, United States 2005-2010"). The study estimated the number of people diagnosed with LD based on medical claims information from a large insurance database. In this study, researchers estimated that 329,000 (range 296,000–376,000) cases of Lyme disease occur annually in the United States.

Results of these studies suggest that the number of people diagnosed with LD each year in the U.S. is around 300,000. Notably, these estimates do not affect our understanding of the geographic distribution of LD. LD cases are concentrated in the Northeast and upper Midwest, with 14 states accounting for over 96% of cases reported to CDC&P. The results obtained using the new estimation methods mirror the geographic distribution of cases that is shown by national surveillance.

2. What is a surveillance case definition?

Reporting of all nationally notifiable diseases, including LD, is based on standard surveillance case definitions developed by the Council of State and Territorial Epidemiologists (CSTE) and CDC&P. The usefulness of public health surveillance data depends on its uniformity, simplicity, and timeliness. Surveillance case definitions establish uniform criteria for disease reporting and should not be used as the sole criteria for establishing clinical diagnoses, determining the standard of care necessary for a particular patient, setting guidelines for quality assurance, or providing standards for reimbursement.

3. How are cases reported to CDC&P?

As with most other reportable diseases, reporting requirements for LD are determined by State laws or regulations. In most states, LD cases are reported by licensed health care providers, diagnostic laboratories, or hospitals. States and the District of Columbia remove all personally identifiable information, then share their data with CDC&P, which compiles and publishes the information for the Nation. CDC&P has no way of linking this information back to the original patient. CDC&P summarizes national surveillance data based on these reports, and publishes results in the CDC&P publication, the "Morbidity and Mortality Weekly Report".

The goal of LD surveillance is not to capture every case, but to systematically gather and analyze public health data in a way that enables public health officials to look for trends and take actions to reduce disease and improve public health.

4. Are more recent numbers available?

Final annual case counts are published when the year is over and all States and territories have verified their data, typically in the fall of the following year. Data from 2016 forward are found in CDC&P WONDER. Selected Lyme disease statistics, tables, and charts are also available on the CDC&P LD website.

CDC&P specific questions

1. What was Lyme Corps?

Lyme Corps was a train-the-trainer program for LD focused on prevention and early recognition of LD and other tick-borne diseases. It ran from 2012 to 2016. Its members consisted of medical and public health students chosen annually from a selected university system in areas of high LD incidence. The members facilitated LD education in conjunction with local health care systems, health departments, schools, and community events (farmers' markets, trail runs, etc.). Lyme Corps participants have also assisted with the following research projects. Universities and states involved in Lyme Corps included:

- Drexel University/Philadelphia Dept. of Health, 2012-2013.

- University of Virginia/ James Madison University, Virginia Department of Health, 2013-2014.

- University of Vermont, Vermont Dept. of Health, 2014-2015.

- University of Maryland/Johns Hopkins University/Uniformed Services University of the Health

Sciences/ Maryland Dept. of Health and Mental Hygiene, 2015-2016.

Lyme Corps members were not Federal employees. Their views and opinions did not necessarily represent the official position of the CDC&P or the U.S. Government.

2. What is CDC&P doing about LD?

CDC&P has a program of service, research, and education focusing on the prevention and control of LD. Activities of this program include:

- Maintaining and analyzing national surveillance data for LD;

- Conducting epidemiological investigations;

- Offering diagnostic and reference laboratory services;

- Developing and testing strategies for the control and prevention of this disease in humans; and

- Supporting education of the public and health care providers.

- In addition, the TickNET program supports research that contributes to the understanding of tick-borne diseases.

3. What are CDC&P specific activities in LD?

- **Leading in Prevention:** Having a long-standing commitment to preventing LD and being on the forefront of disease prevention and control research. CDC&P scientists (entomologists, ecologists, and epidemiologists) have been working together to understand the complicated interactions between ticks, small mammals, deer, and people to help fight this illness. They have worked to find easy, effective, and affordable means for people to fight this illness.

- **Developing better insect repellents:** CDC&P researchers have discovered that a naturally-occurring compound called *nootkatone*, found in grapefruit, Alaska yellow cedar trees, and some herbs, can kill or repel ticks and insects. They are working with an exclusively licensed partner to evaluate and develop next-generation pest prevention and control products.

- **Evaluating permethrin-treated clothing:** CDC&P and university partners are currently evaluating the effectiveness of permethrin-treated clothing as a way to prevent tick bites. Results from a pilot study are now available and additional studies are ongoing.

- **Reducing ticks around homes:** CDC&P and its collaborators have pioneered research in the use of rodent-targeted treatments including bait boxes, a rodent-targeted method of tick control. These devices treat mice and other rodents for ticks, in an effort to upset the transmission cycle of LD. In a 2-year study of 625 homes in Connecticut, scientists examined whether bait boxes reduce the number of tick bites and the incidence of Lyme and other tick-borne diseases.

Results are pending.

- **Exploring rodent-targeted vaccines:** Funding from CDC&P has enabled researchers to develop an oral vaccine for rodents that can reduce the transmission of LD in nature. Evaluation of this product is ongoing.

- **Leading in Surveillance:** Calculating the true cost of LD led CDC&P researchers to the estimate of that over 300,000 people are likely diagnosed and treated each year in the U.S. Additional research is ongoing to determine the economic burden of LD. This research also indicates that the total cost of LD testing alone is estimated at $492 million.

- **Discovering new tick-borne diseases:** CDC&P is partnering with the Minnesota Department of Health, Mayo Clinic, Tennessee Department of Health, and Vanderbilt University to obtain up to 30,000 clinical specimens from patients with suspected tick-borne illness over a 3-year period. CDC&P will use advanced molecular detection to identify tick-borne bacteria that may be the cause of these patients' illnesses. Already, investigators have used AMD to sequence the full genome of a newly discovered bacteria, *Borrelia mayonii*, which is another cause of LD in upper Midwestern states.

- **Improving early and accurate diagnosis:** In collaboration with researchers from Colorado State University, CDC&P scientists are working to develop a new type of test to help healthcare providers diagnose early LD using an innovative approach called "metabolomics". Metabolomics is a type of science that can be used to identify and measure types and amounts of chemicals the body produces during illness. Each type of infection or stage of infection has a different metabolic "fingerprint" that makes it unique. CDC&P is using metabolomics to help develop new testing methods for LD. Early research has shown that metabolic profiling for early LD can be more sensitive than the currently recommended two-tier serology, while retaining high specificity for distinguishing diseases that are not tick-borne (syphilis, severe periodontitis, infectious mononucleosis, and fibromyalgia). New findings from this research could be used to distinguish between LD and Southern Tick-associated Rash Illness (STARI), a condition that occurs after a lone star tick bite but for which there is no known cause.

- Working with Partners: The TickNET program was established by CDC&P in 2007 to bring together expertise from State public health partners and research scientists. TickNET fosters coordinated surveillance, education, and research on the prevention of tick-borne diseases.

- **Funding State Health Departments to improve surveillance and prevention:** CDC&P provides funds to State Health Departments for Lyme and tick-borne disease surveillance through the "Epidemiology and Laboratory Capacity for Infectious Diseases" (ELCID). Coupled with CDC&P's subject matter expertise, this program helps State public health departments strengthen their ability to detect, respond to, control, and prevent LD and other tick-borne diseases.

- **Supporting vector-borne disease Centers of Excellence:** CDC&P has awarded nearly $50 million to five universities to establish regional Centers of Excellence to help effectively address emerging vector-borne diseases in the U.S.. Scientists and public health experts at the Northern and Midwest Regional Centers of Excellence will have a strong research component involving the surveillance and control of disease-carrying ticks.

- **Supporting large-scale prevention research:** CDC&P is contributing both funding and technical expertise to the Tick Project, a 5-year study to determine whether neighborhood-based interventions can reduce cases of Lyme and other tick-borne diseases in people. The study will determine whether two tick control methods, used separately or together, can reduce the number of cases of LD at the neighborhood level. This large-scale research will involve 24 neighborhoods and over 1,000 households.

- **Leading in Public Engagement - Educating healthcare providers:** CDC&P provides many current resources for healthcare providers who wish to learn more about tick-borne diseases, including continuing medical education and guidance in the diagnosis and treatment of tick-borne diseases.

- **Educating the public:** CDC&P has developed and distributed a variety of educational materials ranging from trail signs to patient information sheets–that are free to download or order. CDC&P's LD web site reaches over 7 million people annually and its distribution center has sent out over 30,000 trail signs, 350,000 brochures, and 400,000 prevention bookmarks in the last 5 years. CDC&P's annual Health & Human Services Special Webinars on Lyme and Tick-borne Diseases in collaboration with the (U.S.) FDA and the (U.S.) NIH have reached over 45,000 viewers since they began.

17

What about the other diseases transmitted by the Lyme tick vector?

As seen in Chapter 2, Table 2.1, the black-legged tick *Ixodes scapularis* transmits several other pathogens in addition to *Borrelia burgdorferi,* which is responsible for LD. Table 17.1, an extract of Table 2.1 (line 2) lists these several other diseases: *Anaplasmosis, Ehrlichiosis, Babesiosis,* and Powassan virus disease.

Table 17.1 – Diseases transmitted by the black-legged tick *Ixodes scapularis*

Tick common name	Vector	Bacterium	Disease(s) caused	Where found in the U.S.?	High-risk seasons
Black-legged	*Ixodes scapularis*	o ***Borrelia burgdorferi*** o ***Borrelia mayonii*** o *Anaplasma phagocytophilum* o ***Borrelia miyamotoi*** o *Ehrlichia chaffensis, Ehrlichia ewingii, Ehrlichia muris eauclairensis* o *Babesia microti* o *Powassan virus*	o *Lyme disease* o *Lyme disease* (newly discovered form) o *Anaplasmosis* o A form of relapsing fever o *Ehrlichiosis* o *Babesiosis* o *Powassan virus disease*	East, upper Midwest, and mid-Atlantic	Spring, Summer, Fall (any time when temperatures are above freezing). All tick life stages bite humans, but lymphs and adults are most commonly found on people

Because of their importance as co-infective and confounding diseases, it would be appropriate to briefly review these other diseases. This is the purpose of this Chapter.

Anaplasmosis

Lyme disease

Anaplasmosis is a disease caused by the bacterium *Anaplasma phagocytophilum*. This organism was previously known by other names, including *Ehrlichia equi* and *Ehrlichia phagocytophilum*, and the disease was also previously known as *human granulocytic ehrlichiosis* (HGE).

However, a taxonomic change in 2001 identified that this organism belonged to the genus Anaplasma, resulting in a change in the name of the disease to *anaplasmosis*.

In the United States, *anaplasmosis* was first recognized as a human disease in the mid-1990s, but did not become nationally notifiable until 1999.

What is the epidemiology of *anaplasmosis*?

Since the disease became reportable, the estimated total number of *anaplasmosis* cases reported to CDC&P has increased steadily from 351 cases in 2000, to 5,762 in 2017. In 2000, its has also increased from 1.4 cases per million persons to 17.9 cases per million persons in 2017. The case fatality rate (that is, the proportion of *anaplasmosis* patients that reportedly died as a result of the infection) has remained low, at less than 1%.

Table 17.2 – Number of anaplasmosis cases reported (2000-2017)

Year of report	Number of cases	Year of report	Number of cases
2000	351	2000	1.161
2001	261	2010	1.761
2002	511	2011	2.575
2003	362	2012	2.389
2004	537	2013	2.782
2005	786	2014	2,800
2006	646	2015	3.656
2007	834	2016	4,151
2008	1,009	2017	5,762

Source: CDC&P

Table 17.2 is the number of *anaplasmosis* cases annually reported to the CDC&P during the time period 2000 a 2017. These numbers are also charted in Figure 17.1. It is readily seen that cases of *anaplasmosis* have generally increased from 350 cases in 2000, when the disease became nationally notifiable, to 1,163 cases in 2009, and 5,762 cases in 2017. The number of cases increased 39% between 2016 and 2017.

Seasonality

Although cases of *anaplasmosis* can occur during any month of the year, the majority of cases reported

Lyme disease

to the CDC&P have an illness onset during the summer months and a peak in cases typically occurring in June and July (see Figure 17.2).

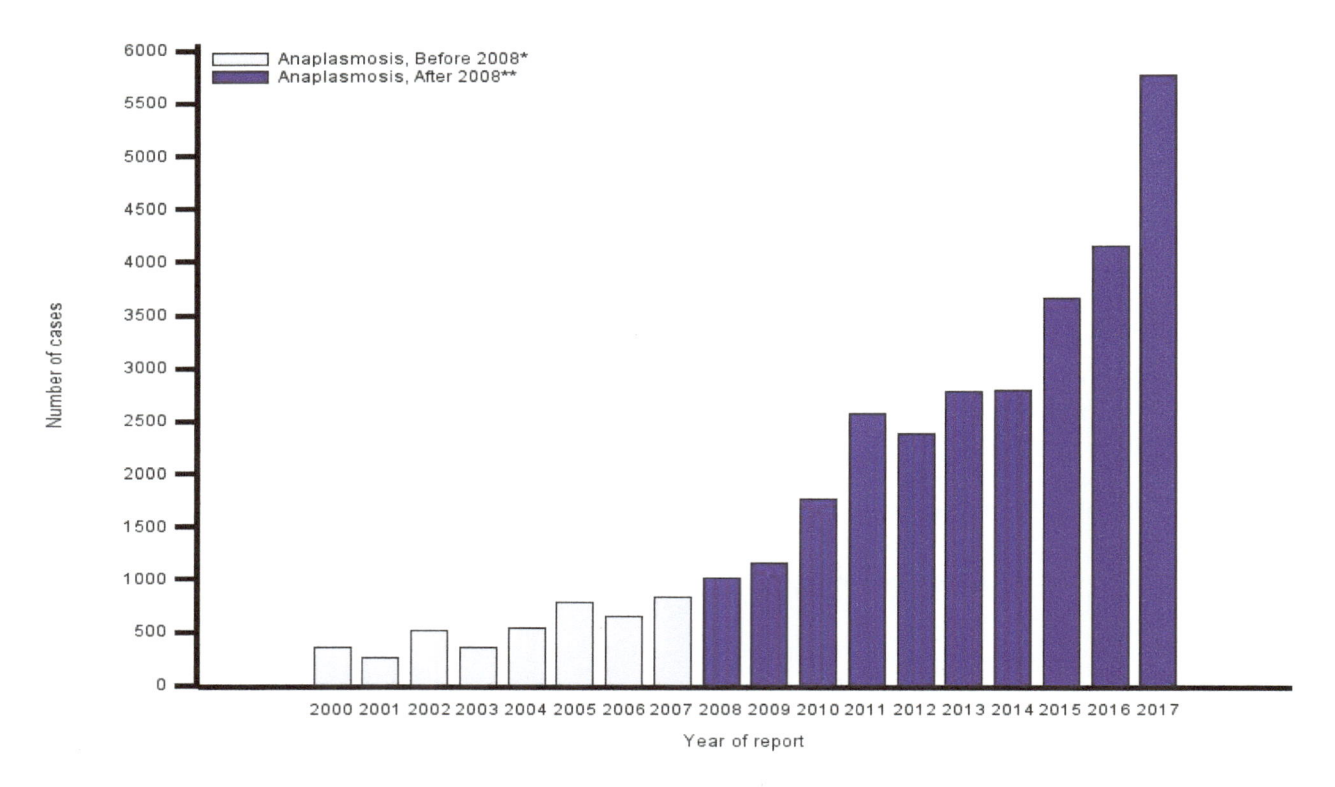

Figure 17.1 - Number of U.S. *anaplasmosis* cases reported to CDC&P (2000–2017)

Source: CDC&P

This period is the season for increased numbers of nymphal black-legged ticks, which is the primary life stage of this tick that bites humans and can transmit the pathogen. A second, smaller peak occurs in October and November and corresponds with the period of adult black-legged tick activity. Figure 17.2 shows the number of cases reported from 2000 through 2017 by month of onset to give the seasonality of cases. There are cases reported in each month of the year, however most are reported in June and July. More than 50% of all cases occur in June and July.

Geographical distribution

Anaplasmosis is most frequently reported from the upper midwestern and northeastern United States. These areas correspond with the known geographic distribution of the black-legged tick (*Ixodes scapularis*), the primary tick vector of *Anaplasma phagocytophilum*. In addition to *Borrelia burgdorferi*, the LD agent, this tick also transmits other human pathogens and co-infections with these organisms (as listed in Table 17.1) have occasionally been reported.

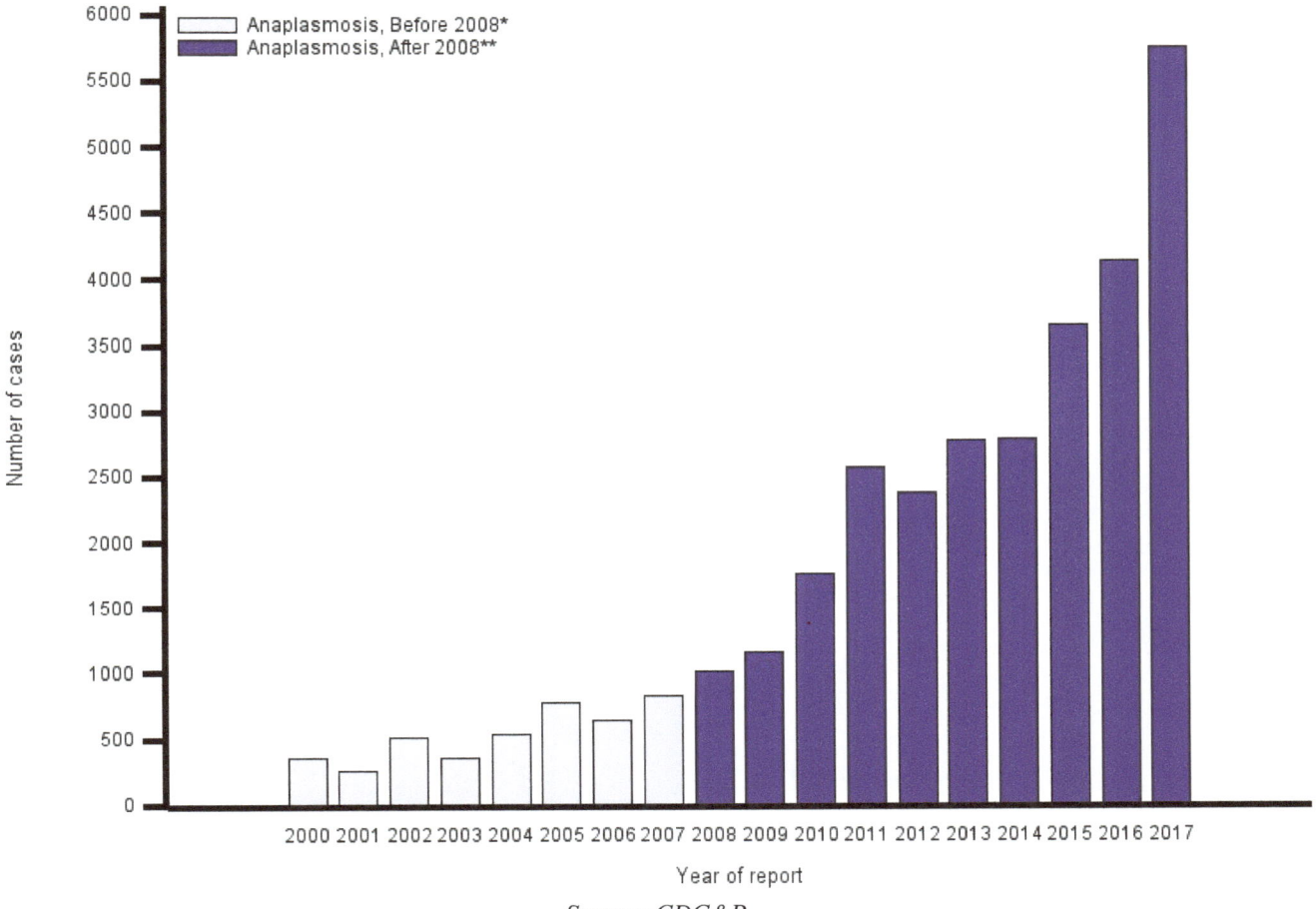

Figure 17.2 - Number of reported U.S. *anaplasmosis* cases by month of onset (2000–2017)

Source: CDC&P

The geographical ranges of *anaplasmosis* (see Figure 17.3) appear to be increasing, which is consistent with the black-legged tick's expanding ranges that have been documented along the Hudson River Valley, Michigan, and Virginia. Eight states (Maine, Massachusetts, Minnesota, New Hampshire, New York, Rhode Island, Vermont, and Wisconsin,) account for 90% of all reported cases of *anaplasmosis*. Additionally and occasionally, *anaplasmosis* cases are reported in other parts of the U.S., including southeastern and south-central States where the organism has not been commonly found. Some of these cases might be due to patient travel to States with higher levels of disease, or misdiagnosis of *anaplasmosis* in patients actually infected with another closely related tick-borne disease (*ehrlichiosis*).

For the same year 2017, Table 17.3 lists the incidence rates by State of *anaplasmosis* per million individuals.

Lyme disease

Table 17.3 – Incidence rates of *anaplasmosis* by State (2017)

State	Incidence rate	State	Incidence rate
o Alaska o Colorado o District of Columbia o Hawaii o Idaho o New Mexico	Not notifiable	o Arizona o Florida o Illinois o Kansas o Kentucky o North Carolina o Ohio o Oklahoma	0.2 – 1.1
o Georgia o Indiana o Louisiana o Mississippi o Nebraska o Nevada o South Carolina o Wyoming	0	o South Dakota o West Virginia o Alabama o Arkansas o Delaware o Maryland o Michigan o Missouri o Montana o Pennsylvania o Tennessee o Virginia	1.1 – 9.4
o California o Oregon o Texas o Washington	> 0 – 0.2	o Connecticut o Maine o Massachusetts o Minnesota o New Hampshire o New Jersey o New York o North Dakota o Rhode Island o Vermont o Wisconsin	> 9.4

Source: Data from CDC&P

People at risk

The frequency of reported cases of *anaplasmosis* is highest among:

- Males and people over 40 years of age;

Lyme disease

- People with weakened immune systems (such as those occurring due to cancer treatments, advanced HIV infection, prior organ transplants, or some medications); and

- People who live near or spend time in known tick habitats.

It behooves everyone to be aware of the degree of risk of getting *anaplasmosis* incurred when living in, or visiting, any of the States listed above.

Figure 17.3 – Geographical distribution in the U.S. of *anaplasmosis* cases

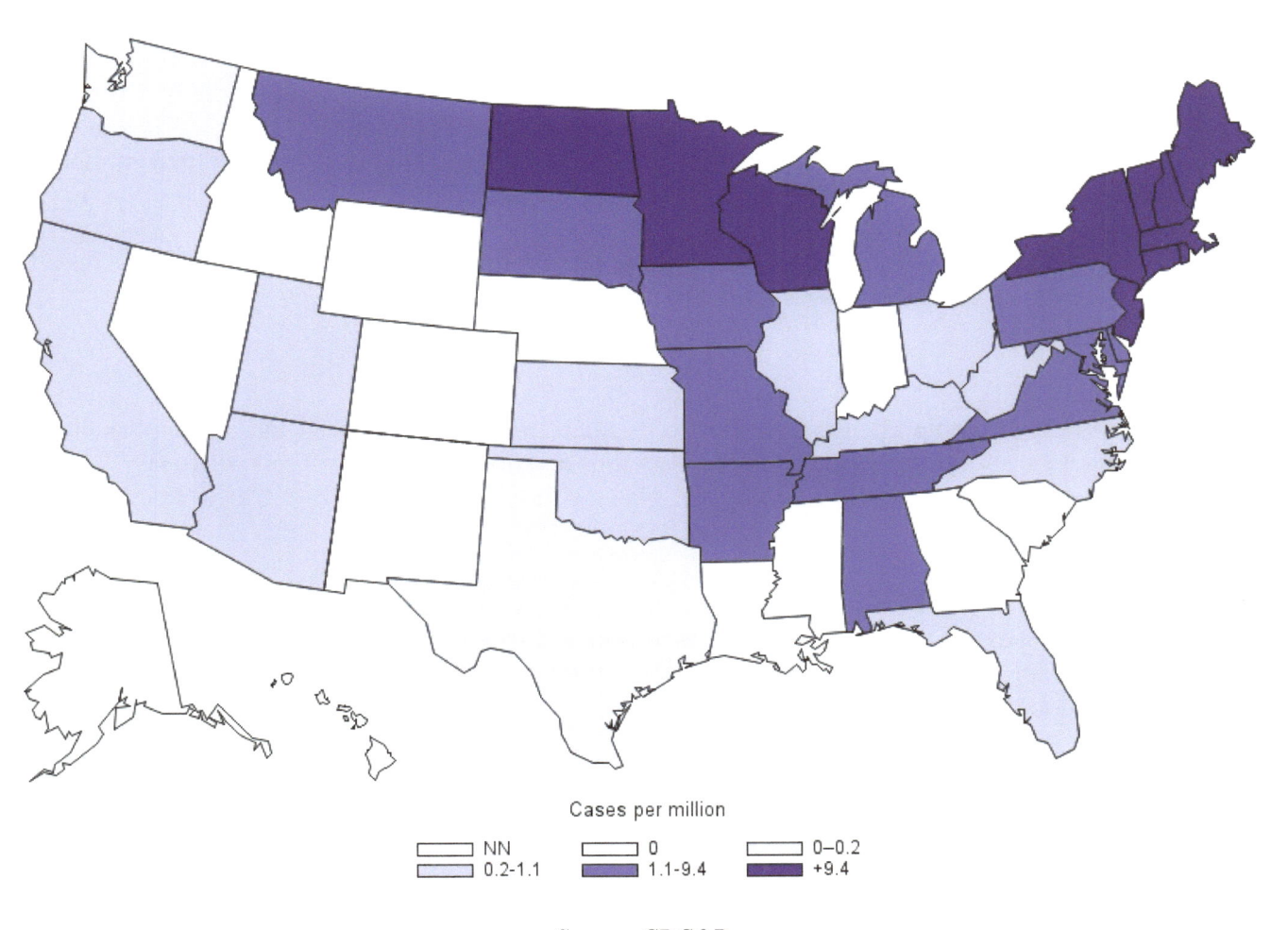

Cases per million

NN | 0 | 0–0.2
0.2–1.1 | 1.1–9.4 | +9.4

Source: CDC&P

Borrelia miyamotoi disease

Disseminated by the agent *Borrelia myamotoi,* and sometimes called hard tick relapsing fever, this disease has been reported as the cause of human infection in the Upper Midwest, the Northeast, and the mid-Atlantic States, in places where LD occurs. However, unlike LD, which is most common in June and July, *Borrelia miyamotoi* infection occurs most commonly in July and August and may be spread by larval black-legged ticks. The incubation period ranges from days to weeks, but specific ranges are unknown.

Signs and symptoms

Its signs and symptoms include: abdominal pain, anorexia (uncommon), arthralgia, chills, confusion, diarrhea, dizziness, dyspnea (uncommon), fatigue, fever, headache (severe), myalgia, nausea, rash (uncommon), and vertigo (uncommon).

Diagnosis

Based on general laboratory findings (elevated hepatic transaminase values, leukopenia, and thrombocytopenia), the laboratory diagnosis relies on polymerase chain reaction (PCR) tests that detect DNA from the organism; or antibody-based tests. The tests are available from a limited number of CLIA-approved reference laboratories.

Recent studies indicate that the C6 peptide ELISA test (a first-tier test for Lyme disease) may be positive in patients infected with *Borrelia miyamotoi*.

Treatment

To date, there are no comprehensive studies to evaluate treatment regimens but, in published case series, patients were successfully treated with antibiotics and dosages used for Lyme disease.

Ehrlichiosis

Ehrlichiosis is the general name used to describe diseases caused by the bacteria *Ehrlichia chaffensis*, *Ehrlichia ewingii*, or *Ehrlichia muris eauclairensis* in the United States. The majority of reported cases are due to infection by *Ehrlichia chaffensis*.

How is *Ehrlichiosis* transmitted?

The majority of *Ehrlichiosis* cases reported to CDC&P have an illness onset during the summer months with a peak in cases typically occurring in June and July. The disease is transmitted in two major ways by:

- **Tick bites:** Most people get *Ehrlichiosis* from the bite of an infected tick, *Ehrlichia chaffensis* and *Ehrlichia ewingii* are transmitted by the lone star tick *Amblyomma americanum* (see Chapter 11 on STARI), found primarily in the south-central and eastern U.S. *Ehrlichia muris eauclairensis* is spread by the black-legged tick *Ixodes scapularis*, which is widely distributed in the eastern United States although cases have only been reported in transmission but it does not eliminate it.

- ***Organ transplantation:*** Two instances of *Ehrlichia chaffensis* transmission through renal transplant from a common donor have been reported. Patients who develop *Ehrlichiosis* within a month of receiving a blood transfusion or solid organ transplant should be reported to State health officials for prompt investigation. (Note: For more information on *Ehrlichiosis* and blood transfusions see the Special Considerations section of the publication: "Diagnosis and Management of Tick-borne Rickettsial Diseases: Rocky Mountain Spotted fever and Other Spotted Fever Group *Rickettsiosis, Ehrlichiosis* and *Anaplasmosis* – United States: A Physical Guide for Health Care and Public Health Professionals (2016)".

Geographical distribution

It is illustrated in Figure 17.4 below.

Figure 17.4 – Estimated geographical distribution of the Lone Star tick in the U.S.

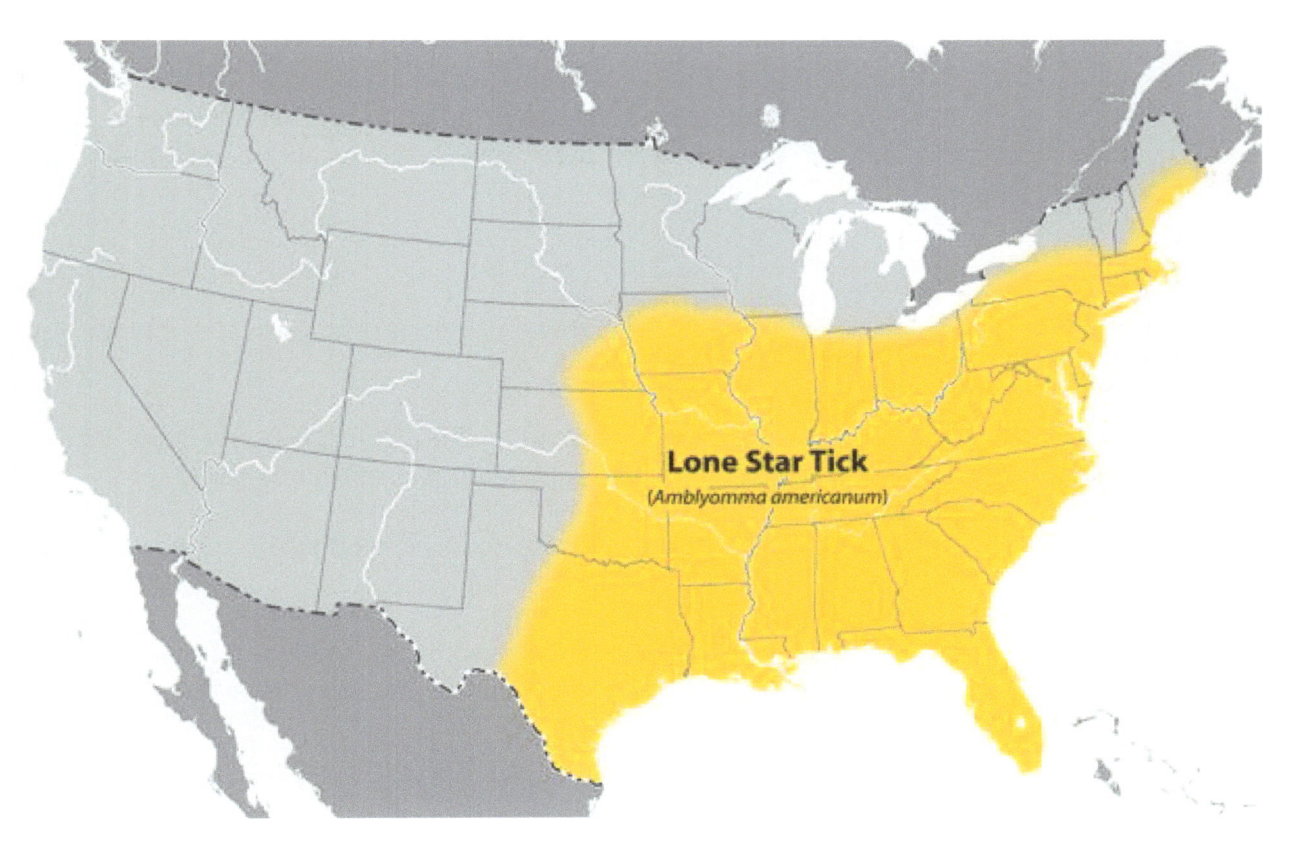

Source: CDC&P

Epidemiology

The geographic range of *Ehrlichiosis* cases depends highly on the species of ehrlichia-causing illness:

Ehrlichia chaffensis and *Ehrlichia ewingii* infections occur primarily in south-central, southeastern, and mid-Atlantic states. *Ehrlichiosis muris eauclairensis* infections have only been reported from Wisconsin and Minnesota and travelers to those States.

Babesiosis

Babesiosis is caused by parasites that infect red blood cells (RBC). Most U.S. cases are caused by *Borrelia microti* (and other *babesia* species), which is transmitted by *Ixodes scapularis* ticks, primarily in the Northeast and Upper Midwest. *Babesia* parasites also can be transmitted via transfusion, anywhere, at any time of the year. The incubation period is 1–9+ weeks.

Babesia infection can range from asymptomatic to life threatening. Risk factors for severe *babesiosis* include asplenia, advanced age, and impaired immune function. Severe cases can be associated with marked thrombocytopenia, disseminated intravascular coagulation, hemodynamic instability, acute respiratory distress, renal failure, hepatic compromise, altered mental status, and death.

Geographical distribution

Babesiosis is most frequently reported from the northeastern and Upper Midwestern United States in areas where *Borrelia microti* is endemic. Sporadic cases of infection caused by novel *babesia* agents have been detected in other U.S. regions, including the West Coast. In addition, transfusion-associated cases of *babesiosis* can occur anywhere in the country.

Signs and Symptoms

The signs and symptoms include: arthralgia, chill, fatigue, fever, gastrointestinal symptoms (anorexia, nausea); headache, malaise, myalgia, sweats, dark urine.

Less common symptoms include: conjunctival infection, cough, depression, emotional lability, other gastrointestinal symptoms (abdominal pain, vomiting), photophobia, and sore throat.

Mild symptoms may occur in some patients and include: mild hepatomegaly, mild splenomegaly, or jaundice.

Not all infected persons are symptomatic or febrile. The clinical manifestations, if any, usually develop within several weeks after exposure, but may develop or recur months later (for example, in the context of surgical splenectomy).

Diagnosis

In addition to signs and symptoms, the diagnosis is based on general laboratory findings including: decreased hematocrit due to hemolytic anemia, thrombocytopenia; elevated serum creatinine and blood urea nitrogen (BUN) values, and mildly elevated hepatic transaminase values.

The laboratory diagnosis includes: identification of intra-erythrocytic *babesia* parasites by light-microscopic examination of a peripheral blood smear; or positive PCR analysis for *Babesia microti*; or isolation of *babesia* parasites from a whole blood specimen by animal inoculation (in a reference laboratory).

Note that sometimes it can be difficult to distinguish between *babesia* and malaria parasites and even between parasites and artifacts (such as stain or platelet debris).

Treatment

In March 2018, the FDA approved the first *Borrelia microti* blood donor screening tests. Congenital transmission has also been reported. Treatment decisions and regimens should consider the patient's age, clinical status, immune-competence, splenic function, co-morbidities, pregnancy status, other medications, and allergies. Expert consultation is recommended for persons who have or are at risk for severe or relapsing infection or who are at either extreme of age.

For ill patients, *babesiosis* usually is treated for at least 7–10 days with a combination of two medications, typically either (atovaquone + azithromycin) or (clindamycin + quinine: the standard of care for severely ill patients).

Powassan virus disease

Powassan virus infections have been recognized in the United States, Canada, and Russia. In the U.S., cases have been reported primarily from Northeastern States and the Great Lakes region. The causing agent is the Powassan virus with an incubation period of 1–4 weeks.

Signs and Symptoms

Signs and symptoms include: fever, headache, vomiting, and generalized weakness. The illness usually progresses to meningo encephalitis, which may include: aphasia, cranial nerve palsies, meningeal signs, altered mental status, or movement disorders, paresis, or seizures.

Diagnosis

In addition to the signs and symptoms, the diagnosis is based on laboratory findings including: CSF findings (lymphocytic pleocytosis), normal or mildly elevated protein, and normal glucose. The laboratory diagnosis includes primarily testing available at CDC&P and selected State Health Departments, and limited commercial testing; and measurement of virus-specific IgM antibodies in serum or CSF.

Treatment

No specific antiviral treatment for Powassan virus disease is available. Patients with suspected Powassan virus disease should receive supportive care as appropriate.

Take-home points

- The black-legged tick *Ixodes scapularis* transmits several other pathogens in addition to *Borrelia burgdorferi,* which is responsible for Lyme Disease including *anaplasmosis, ehrlichiosis, babesiosis,* and Powassan virus disease. They are important as co-infective and confounding diseases for Lyme disease.

- *Anaplasmosis* is a disease caused by the bacterium *Anaplasma phagocytophilum*. The estimated total number of anaplasmosis cases reported to CDC&P has increased steadily.

- Although cases of *anaplasmosis* can occur during any month of the year, the majority of cases reported to the CDC&P have an illness onset during the summer months and a peak in cases typically occurs in June and July. *Anaplasmosis* is most frequently reported from the upper midwestern and northeastern United States. It corresponds with the known geographic distribution of the black-legged tick.

- In addition to *Borrelia burgdorferi,* the Lyme disease agent, this black-legged tick tick also transmits other human pathogens and co-infections with these organisms.

- People at risk for *anaplasmosis* are males and people over 40 years of age; people with weakened immune systems; and people who live near or spend time in known tick habitats.

- Disseminated by the agent *Borrelia myamotoi,* and sometimes called hard tick relapsing fever, *borreliosis* has been reported as the cause of human infection in the Upper Midwest, the Northeast, and the mid-Atlantic States, in places where Lyme disease occurs.

- The signs and symptoms of *borreliosis* include: abdominal pain, anorexia (uncommon), arthralgia, chills, confusion, diarrhea, dizziness, dyspnea (uncommon), fatigue, fever, headache (severe), myalgia, nausea, rash (uncommon), and vertigo (uncommon). To date, there are no comprehensive studies to evaluate treatment regimens for *borreliosis*, but patients were successfully treated with antibiotics at dosages used for Lyme disease.

- *Ehrlichiosis* is caused by the bacteria *Ehrlichia chaffensis, Ehrlichia ewingii,* or *Ehrlichia muris eauclairensis* in the United States. The majority of reported cases are due to infection by *Ehrlichia chaffensis*. The illness onset is during the summer months with a peak in cases typically occurring in June and July, primarily in south-central, southeastern, and mid-Atlantic states. They have only been reported from Wisconsin and Minnesota and by travelers to those States. The disease is transmitted by tick bites or blood transfusion or/and organ transplant.

- *Babesiosis* is caused by parasites that infect red blood cells. Most U.S. cases are caused by *Borrelia microti* (and other *babesia species*), which is transmitted by *Ixodes scapularis* ticks, primarily in the Northeast and Upper Midwest. The parasites can be transmitted via transfusion, anywhere, at any time of the year. The incubation period is 1–9+ weeks. *Babesia* infection can range from asymptomatic to life threatening. Sometimes it can be difficult to distinguish between *babesia* and malaria parasites.

- Treatment decisions and regimens for *babesiosis* should consider the patient's age, clinical status, immune-competence, splenic function, co-morbidities, pregnancy status, other medications, and allergies. Expert consultation is recommended for persons who have or are at risk for severe or relapsing infection or who are at either extreme of age. For ill patients, *babesiosis* is usually treated for at least 7–10 days with a combination of two medications, typically either (atovaquone + azithromycin) or (clindamycin + quinine: the standard of care for severely ill patients).

- Powassan virus infections have been recognized in the United States, Canada, and Russia. In the U.S., cases have been reported primarily from Northeastern States and the Great Lakes region. The causing agent is the Powassan virus with an incubation period of 1–4 weeks. Signs and symptoms include: fever, headache, vomiting, and generalized weakness. The illness usually progresses to meningo-encephalitis, which may include aphasia, cranial nerve palsies, meningeal signs, altered mental status, movement disorders, paresis, or seizures. No specific antiviral treatment for Powassan virus disease is available. Patients with suspected Powassan virus disease should receive supportive care as appropriate.

- It behooves everyone to be aware of the degree of risk of getting any of the above infectious diseases when living in, or visiting, any of the States correspondingly listed.

18

More about other tick-borne diseases of the U.S.

In addition to Sidebar 2.1 and Table 2.2, I provide in this Chapter further details on the tick-borne associated diseases of the United States. A good reference is provided by a publication of the (U.S.) Department of Health & Human Services, Center for Disease Control & Prevention (CDC&P) titled "Tick-borne diseases of the United States - A reference manual for health care providers" (2018). I have reproduced its cover page in Figure 18.1 below.

Although addressed to the health care provider, that publication is generally understandable and could be of interest to others, particularly those afflicted by Lyme, in several regards. It systematically considers in a succinct fashion each of the ticks of interest, providing its biological identification, the geographical area(s) where it is found, its signs and symptoms, its mode(s) of transmission, the corresponding laboratory tests and their diagnostic value, and its treatment (if any is available). The several tick-borne diseases considered are:

- *Anaplasmosis*;

- *Borrelia miyamotoi* **disease**;

- *Ehrlichiosis*;

- *Babesiosis*;

- **Powassan virus disease**;

- **Lyme disease** (also considered at length in this volume, any pertinent additional information or illustration is nevertheless added here);

- **Spotted fever** *rickettsiosis* **(SFR)**, including

- **Rocky Mountain spotted fever (RMSF)**; and

- *Tularemia.*

Figure 18.1 – A reference manual for tick-borne diseases of the United States

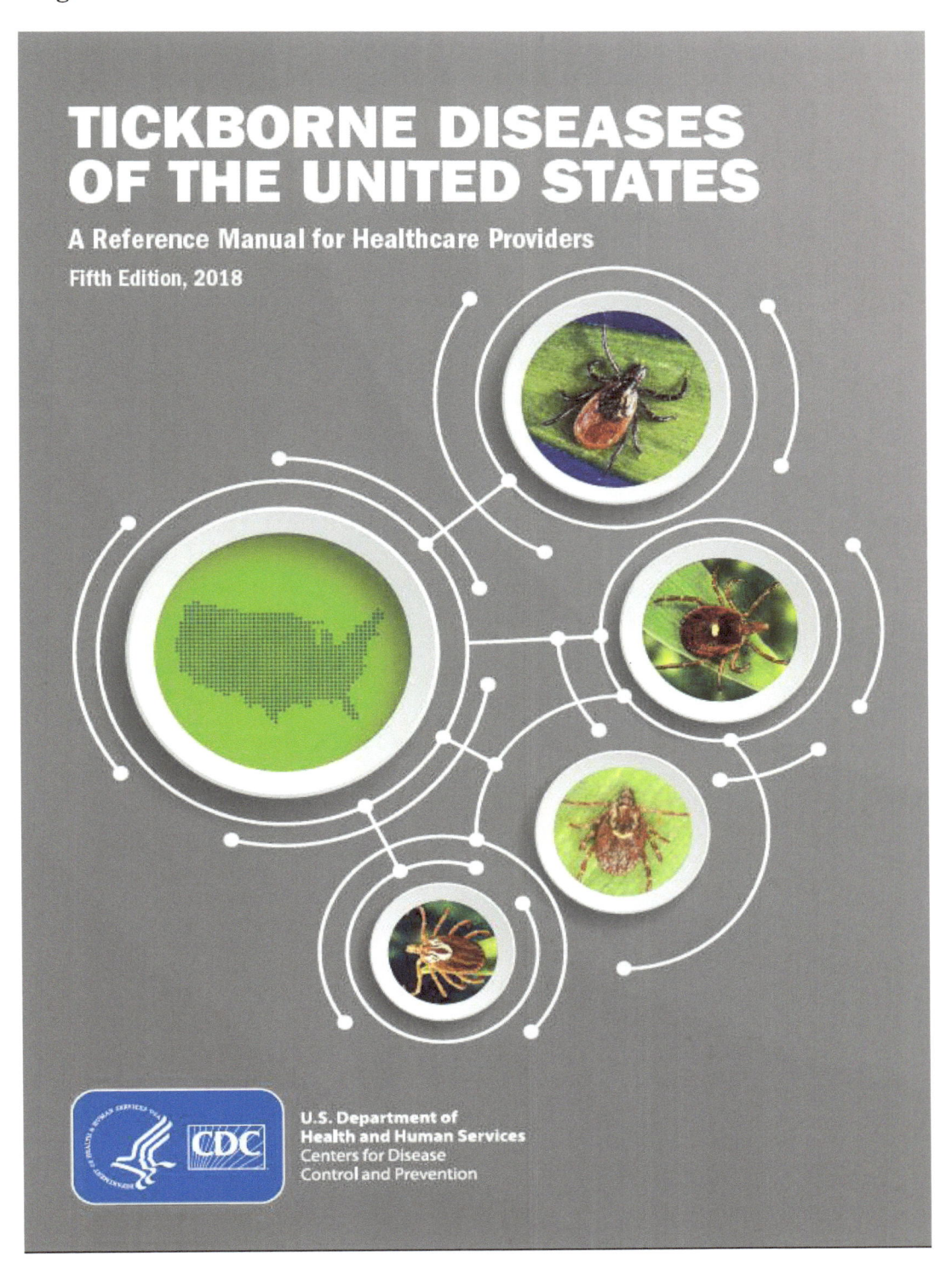

(The other ticks uniquely encountered in Europe and elsewhere are not considered here for this would exceed the scope of this volume.)

The characteristics of all such diseases are summarized in Table 18.1 including, for each agent, where geographically found in the U.S., the incubation period, the signs and symptoms, and the corresponding treatment (if any available).

Table 18.1 – Characteristics of tick-borne diseases of the United States

Tick-borne disease	Bacterium	Where found in the U.S.?	Incubation period	Signs & symptoms	Treatment
Anaplasmosis	*Anaplasma phagocytophilum*	Upper midwest and Northeastern States	14 days	o Fever, chills, severe headache, malaise o Gastrointestinal symptoms (nausea, vomiting, diarrhea, anorexia) o Rash, etc. (< 10%)	o Clinical suspicion is sufficient to begin treatment o Delay in treatment may result in severe illness and even death o Antibiotics (doxycycline)
Babesia	*Babesia microti* and other *babesia* species	Northeastern and midwestern States	1–9+ weeks	o Arthralgia, chills, fatigue, fever, malaise, myalgia, sweats o Gastrointestinal symptoms (nausea, less commonly abdominal pain, vomiting, dark urine) o Less commonly: conjunctival infection, cough, depression, emotional lability, photophobia, sore throat o Mild symptoms: hepatomegaly, jaundice, splenomegaly	Either (tovaquone + azithromycin) or (clindamycin + quinine) for 7- days
Borrelia miyamotoi	*Borrelia miyamotoi*	Upper midwest, northeast, and mid-Atlantic States	Days to weeks (specific ranges unknown)	o Arthralgia/ myalgia, chills, confusion, diarrhea, dizziness, fatigue,	Antibiotics at dose used for Lyme disease

Lyme disease

				fever, severe headache, nausea, abdominal pain, o Uncommonly: anorexia, dyspnea, rash, vertigo	
Colorado tick fever virus (CTFV)	Colorado tick fever virus	Primarily Colorado, Utah, Montana, and Wyoming	1–14 days	o Chills, fever, headache, conjunctival infection, lethargy, myalgia, pharyngeal erythema, and lymphadenopathy o Prolonged convalescence (fatigue in adults, weakness,) o Rare: life threatening complications associated with disseminated intravascular coagulation and/or meningo-encephalitis in children o < 20% of cases: maculopapular or petechial rash o ~ 50% of patients have a biphasic illness	None specific
Ehrlichiosis	*Ehrlichia saffensis* *Ehrlichia ewingii* *Ehrlichia muris eauclairensis*	Southeast and south central States from east coast in Texas	5–14 days	o Chills, fever, headache, malaise, muscle pain o Gastrointestinal symptoms: anorexia, diarrhea, nausea, vomiting, o Altered mental status o Rash (among children)	o Antibiotics (doxycycline) o Delay in treatment may result in severe illness or death
Heartland & Bourbon virus disease (H&BVD)	Heartland virus	Midwest and south States	Unknown (reportedly 2 weeks prior to illness)	o Decreased appetite, arthralgia, diarrhea, fatigue, fever, headache, myalgia, nausea	o None available o Supportive care
Lyme disease	*Borrelia*	o Upper	3–30 days	o *Erythema*	Antibiotics

Lyme disease

	burgdorferi *Borrelia miyamotoi*	midwestern and northeastern States, northern California, Oregon, Washington o In 2015: 95% of cases in 14 States: Connecticut, Delaware, Maine, Maryland, Massachusetts, Minnesota, New Hampshire, New Jersey, Mew York, Pennsylvania, Rhode Island, Vermont, Virginia, Wisconsin		*migrans* (EM) rash o Flu-like symptoms (arthralgia, headache, fever, malaise, myalgia) o Lymph-adenopathy	(different regimens for adults and children): doxycycline, cefuroxime acetil, amoxicillin
Powassan virus disease	Powassan virus	Northeastern States and the Great Lakes	1-4 weeks	o Fever, headache, vomiting, general weakness, o Usually progresses to meningo-encephalitis. o May include aphasia, meningeal signs, altered mental states, movement disorders, cranial nerve palsies, paresis, seizures	o Sane as H&BVD o None available o Supportive care
Rocky mountain spotted fever (RMSF)	*Rickettsia rickettsii*	o Throughout all 52 States o > 60% of cases in 5 States (North Carolina, Oklahoma, Arkansas, Tennessee, and Missouri)	1-3 days	EARLY 1-4 DAYS: o Edema (around eyes, on back of hands), high fever, severe headache, malaise, myalgia, o Gastro-intestinal symptoms (anorexia, nausea, vomiting) LATE >/= 5 DAYS: o Cerebral edema, coma, altered mental status, respiratory compromise	o Antibiotics: doxycycline o Any clinical suspicion is sufficient to begin treatment o Delay in treatment may result in severe illness and even death

Lyme disease

				(ARDS, pulmonary edema), necrosis (requiring amputation), multi-organ system damage (CNS, renal failure)	
Rickettsiosis	*Rickettsia parkeri*	Southern and mid-Atlantic States	2-10 days	o Almost always associated with an inoculation eschar (ulcerated, necrotic lesion) at the site of the tick attachment o Muscle aches, fever, headache, rash	o Same as RMSF o Antibiotics: doxycycline o Any clinical suspicion is sufficient to begin treatment o Delay in treatment may result in severe illness and even death
Tick-borne relapsing fever (TBRF)	*Borrelia hermsii* *Borrelia turicatae*	14 Western States: Arizona, California, Colorado, Idaho, Kansas, Montana, Nevada, New Mexico, Oklahoma, Oregon, Texas, Utah, Washington, and Wyoming	7 days followed by recurring febrile episodes that last ~ 3 days and are separated by afebrile periods of 7 days	Arthralgia, chills, headache, myalgia, nausea, facial palsy (rarely), vomiting	Antibiotics: tetracycline, erythromycin, ceftriaxone
Tularemia	*Francisella tularensis*	All States except Hawaii	3-5 days (range 1-21 days)	Abdominal pain, anorexia, chest discomfort, chills, cough, diarrhea, fatigue, fever, headache, malaise, myalgia, sore throat, vomiting	Antibiotics: streptomycin, gentamycin, ciproflexacin, doxycycline

Source: Adapted from CDC&P

For all such tick-borne diseases, a geographical map of the U.S. is correspondingly provided (Figures 18.2 through 18.7). In each such map, each dot represents one case. Cases are reported from the infected person's county of residence, not necessarily the place where that person may have been infected. The extent of such inaccurate reporting is unknown but any other reporting method does not seem to be practical.

It may be noted that, in 2016, no cases of tick-borne illness were reported from Hawaii; Alaska reported 6 travel-related cases of Lyme disease and 1 case of *tularemia*; *babesiosis* was reportable in 35 States: Alabama, Arkansas, California, Connecticut, Delaware, Illinois, Indiana, Iowa, Louisiana,

Lyme disease

Kentucky, Maine, Maryland, Massachusetts, Michigan, Minnesota, Missouri, Montana, Nebraska, New

Figure 18.2 – U.S. geographical distribution of *Anaplasmosis*

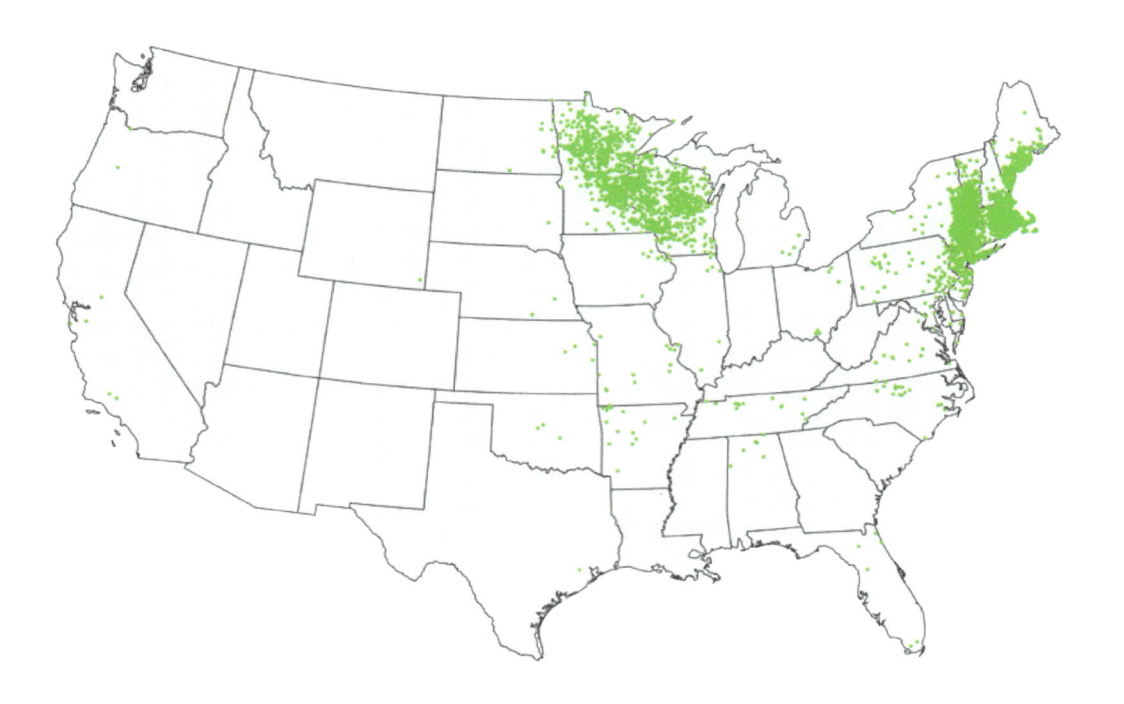

Figure 18.3 – U.S. geographical distribution of *Babesiosis*

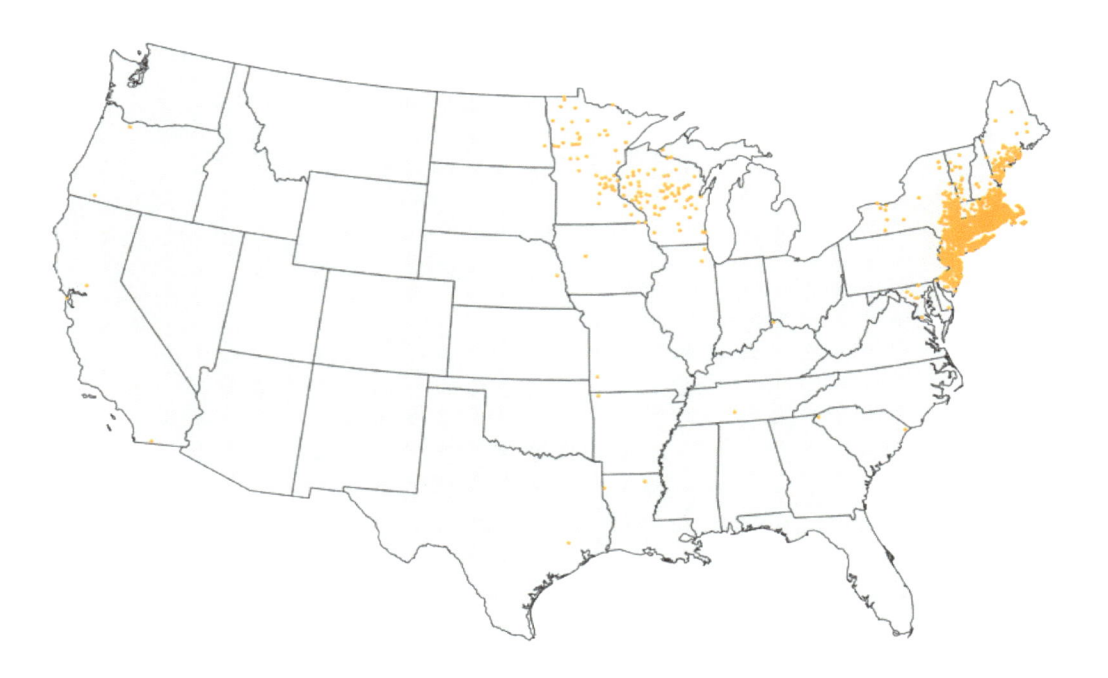

Lyme disease

Figure 18.4 – U.S. geographical distribution of *Ehrlichiosis*

Figure 18.5 – U.S. geographical distribution of Lyme disease

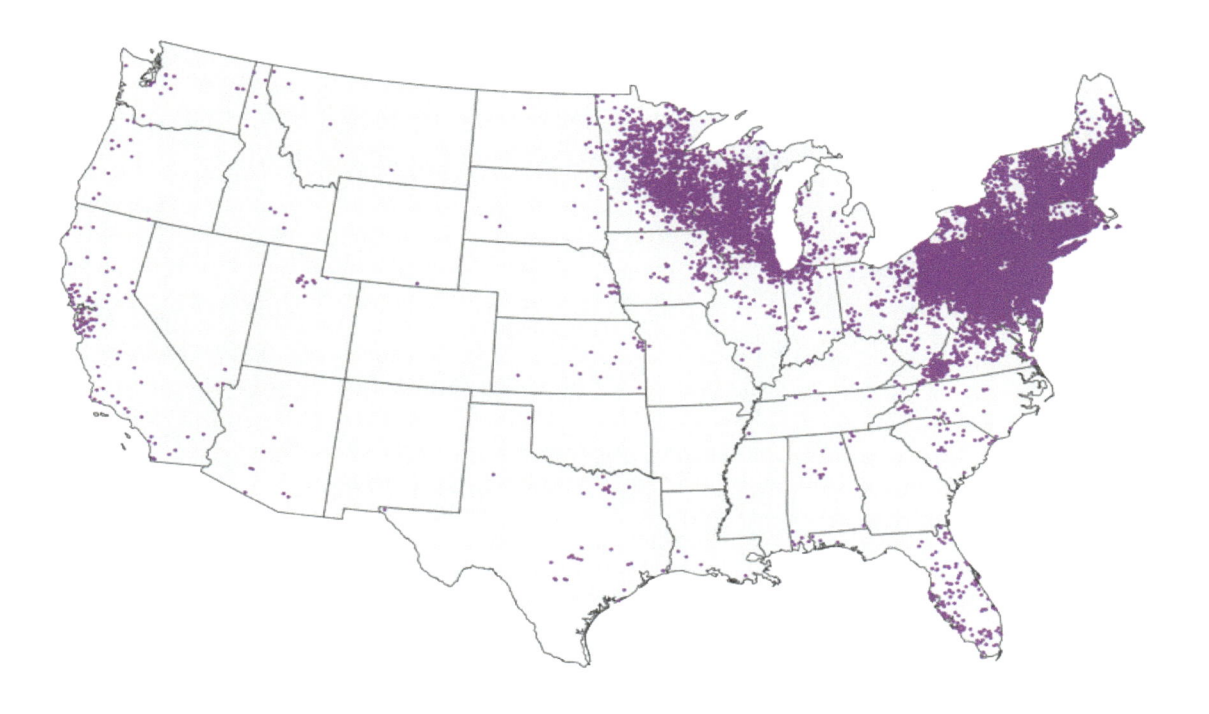

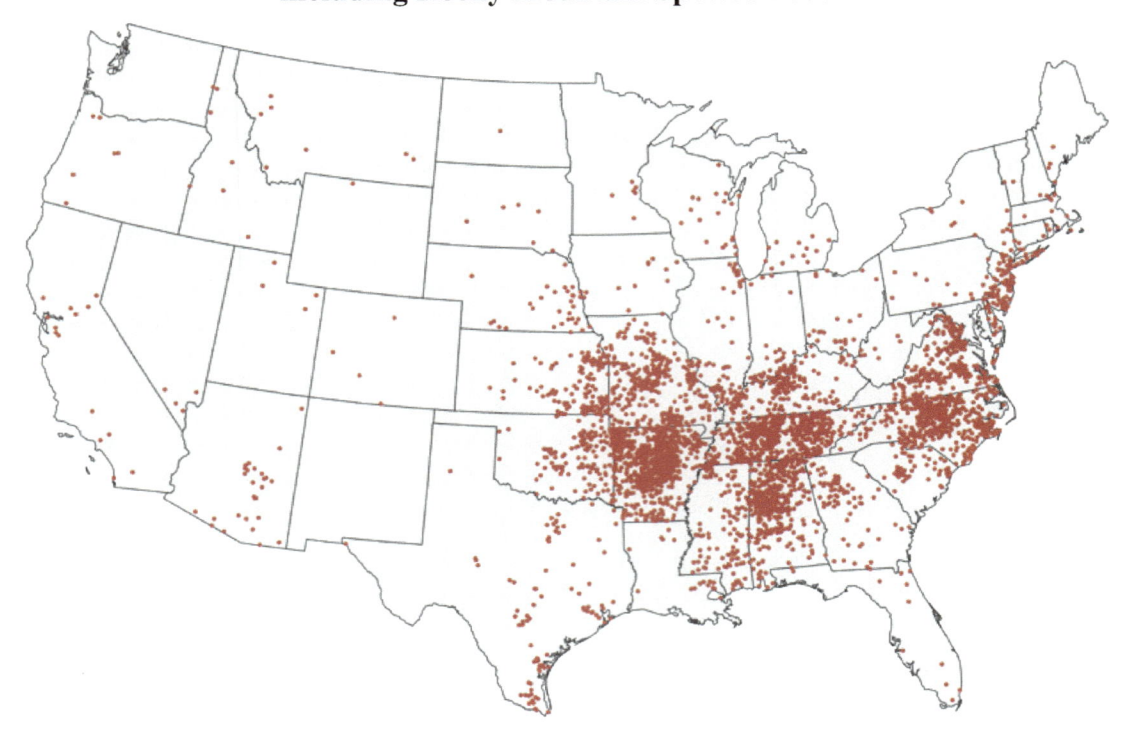

Figure 18.6 – U.S. geographical distribution of Spotted Fever *Rickettsiosis* including Rocky Mountain Spotted Fever

Hampshire, New Jersey, New York, North Dakota, Ohio, Oregon, Rhode Island, South Carolina, South Dakota, Tennessee, Texas, Utah, Vermont, Washington, West Virginia, Wisconsin, and Wyoming.

Also in 2016, *anaplasmosis* and *ehrlichiosis* were not reportable in Colorado, Idaho, New Mexico, Alaska, and Hawaii. Further, spotted fever *rickettsiosis* was not reportable in Alaska and Hawaii.

Take-home points

- Further details on the tick-borne associated diseases of the United States have been provided for *anaplasmosis, babesiosis, ehrlichiosis,* Lyme disease, spotted fever *rickettsiosis* (SFR), including Rocky Mountain spotted fever (RMSF), and *tularemia.*

- The characteristics of all such diseases have been summarized including, for each bacterium, where geographically found in the U.S., the incubation period, the signs and symptoms, and the corresponding treatment (if any available).

- Geographical maps of the U.S. were provided for each of the listed tick-borne diseases to illustrate the regions infected.

Lyme disease

- Case data employed in drafting the above maps are reported from the infected person's county of residence, not necessarily the place of infection. The uncertainty of such inaccurate reporting is unknown.

Figure 18.7 – U.S. geographical distribution of *Tularemia*

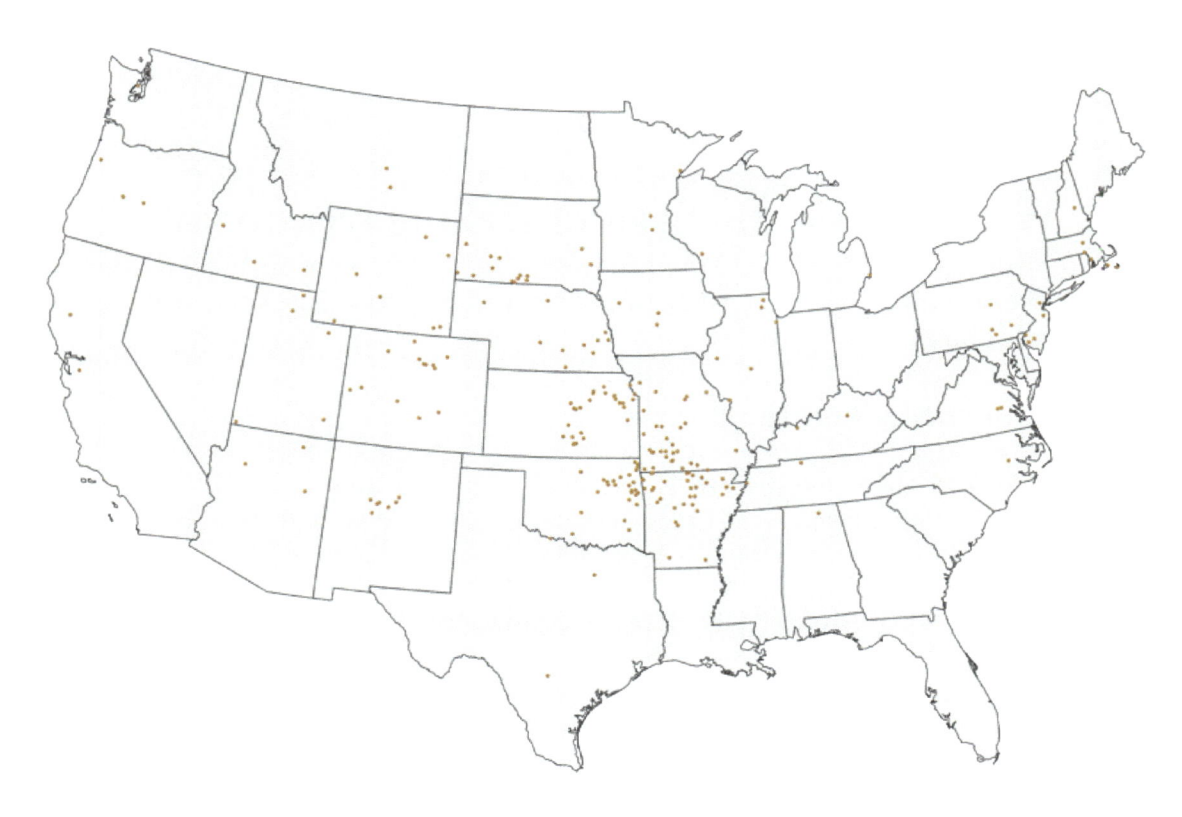

Source: CDC&P

19

What can you do beyond seeking treatment?

Now that you have been diagnosed with LD, what can you do about it beyond seeking treatment? I suggest the following steps:

Become well-informed about the disease

Becoming well-informed about the disease is one important long-term strategy. Also, programs that teach families about the various stages of Lyme and about ways to deal with caregiving challenges can help.

Reach-out to supporting and advocating organizations

Helpful information can be found in the websites of the following supporting and advocating organizations or by contacting them directly:

Infectious Diseases Society of America (IDSA)

The IDSA is a community of over 12,000 physicians, scientists, and public health experts who specialize in infectious diseases. Its purpose is to improve the health of individuals, communities, and society by promoting excellence in patient care, education, research, public health, and prevention relating to infectious diseases. Its website contains a trove of useful information (2,748 results in a search of its website).

Contact:
1300 Wilson Boulevard, #300, Arlington, VA 22209, U.S.A.
Tel: 703-299-020
Website: idsociety.org

Lyme Disease Association (LDA)

The LDA is a U.S. national non-profit education, research, prevention, and patient support organization. Its aim is to promote awareness and control of the spread of Lyme and other tick-borne diseases (TBD), and their complications.

Contact:
PO Box 1438, Jackson, NJ 08527
888-366-6611 | information line
732-938-7215 | fax
email: LDA@LymeDiseaseAssociation.org |

Global Lyme Alliance (GLA)

The GLA has forged significant partnerships with the Lyme Disease Association and the academic research community. It created the first research center for the study of persistent Lyme at Columbia University Medical Center in New York City. For the past decade, it has funded innovative and promising research at Johns Hopkins University, Washington University in St Louis, Texas A&M, Northeastern University, University of Missouri, University of California San Francisco, University of Pennsylvania, and Stony Brook University. It also allied with Ionica Sciences to develop an accurate Lyme diagnostic test. Its website contains helpful information on prevention, diagnosis, and treatment.

Contact:
203-969-1333
email: info@gla.org

European Society of Clinical Microbiology and Infectious Diseases (ESCMID)

The ESCMID is a non-profit organization with mission to improve the diagnosis, treatment, and prevention of infection-related diseases. It supports research, education, training, and good medical practice. It is Europe's leading society in clinical microbiology and infectious diseases with members from all European countries and all continents.

Contact:
POB 214
4010 Basel, Switzerland
Gerbergasse 14, Third Floor
4001 Basel, Switzerland
Tel: +41 61 5080 173
email: info@escmid.org

International Society for Environmentally Acquired Diseases (ISEAD)

The ISEAD is a non-profit professional medical society with aim to raise awareness of the environmental causes of inflammatory illnesses and to support the recovery of individuals affected by these illnesses through the integration of clinical practice, education, and research. Its aim is to restore health to individuals with environmentally acquired illnesses through clinical practice, education, and research.

Contact:

121 S Estes Rd, # 205D
Chapel Hill, North Carolina 27514
Website: isead.org

Lyme Hope, Canada
Especially for maternal-placental transmission of Lyme disease.

Follow at:
twitter@lymehopecanada/advocacy-updates
twitter@lymemoms
https//:facebook.com/lymehopecanada/
https://youtube/SLFRYVcGeR4

Other resources:

Association of State and Territorial Public Health Laboratory Directors (ASTPHLD)

Contact:
Telephone (202) 822-5227

(U.S.) Centers for Disease Control & Prevention (CDC&P)
Division of Vector-borne Diseases, National Center for Emerging and Zoonotic Infectious Diseases

(U.S.) Food & Drug Administration (FDA)

Contact:
Website: www.fda.gov
Telephone: (888) 463-6332

MedlinePlus.gov
National Library of Medicine
www.medlineplus.gov

National Human Genome Research Institute (NHGRI)
www.genome.gov/health

National Center for Biotechnology Information (NCBI)

National Library of Medicine
www.ncbi.nlm.nih.gov

Sjögren Syndrome Foundation (SSF)

Contact:

6707 Democracy Blvd, Suite 325, Bethesda, MD 20817.
www.sjogrens.org
Tel: 800-475-6473
e-mail ssf@sjogrens.org

For neuropathy:

American Academy of Neurology (AAN)
The AAN strives to promote the highest quality patient-centered neurologic care and enhance member career satisfaction

Contact:
201 Chicago Avenue
Minneapolis, MN 55415
Website: www.aan.com
Tel: (800) 879-1960
or (612) 928-6000 (International)
Fax: (612) 454-2746

American Neurological Association (ANA)
The ANA is devoted to advancing the goals of academic neurology, to training and educating neurologists and other physicians in the neurologic sciences, and to expanding both our understanding of diseases of the nervous system and our ability to treat them.

Contact:
1120 Route 73, Suite 2004
10.027Mount Laurel, New Jersey 08054
Website: www.myann.org
Telephone: (856) 638-0423

Society for Neuroscience (SfN)
The SfN is dedicated to advancing the understanding of the brain and the nervous system, providing professional development activities, information, and educational resources for neuroscientists, promoting public information and general education about the latest neuroscience research, supporting active and continuing discussions on ethical issues relating to the conduct and outcomes of neuroscience research, and inform legislators and other policymakers about new scientific knowledge, recent developments, and emerging opportunities in neuroscience research and their implications for public policy, societal benefit, and continued scientific progress.

Contact:
1121 14th Street NW Suite 1010
Washington, D.C. 2005
Website: www.sfn.org
Telephone: (202) 962-4000

Western Neuropathy Association
The WNA is a non-profit, all-volunteer organization committed to assisting and providing hope through caring, support, research, education, and empowerment.

Contact:
POB 276567
Sacramento, California 95827
Website: www.pnhelp.org
Telephone: (877) 622-6298

Foundation for Peripheral Neuropathy (FPN)
The FPN is a public charity foundation committed to fostering collaboration among today's most gifted and dedicated neuroscientists and physicians to help maintain a comprehensive view of the field and determine the research areas that hold the most promise in neuropathy research and treatment with the ultimate goal to dramatically improve the lives of those living with this painful and debilitating disorder.

Contact:
485 Half Day Road Suite 250
Buffalo Grove, Illinois 60089
Website: www.foundationforpn.org
Telephone: (877) 882-9942
Fax: (887) 882-9960

Begin treatment early in the disease process

If you think you may have Lyme disease, contact your physician right away for diagnosis and treatment as discussed at length in Chapters 5-8. The FDA regulates diagnostic tests to ensure that they are safe and effective. It is important to know that blood tests that check for antibodies (produced by the body to fight infection) to the bacterium that causes Lyme disease are not useful if done soon after a tick bite. It typically takes 2-5 weeks after a tick bite for initial antibodies to develop.

According to the CDC&P, patients treated with appropriate antibiotics before the diagnostic tests are complete in the early stages of Lyme disease usually recover rapidly and completely.

Look for new diagnostic tests

In addition to the tests described and discussed in Chapter 5, new approaches to diagnosis are being pursued. Keep abreast of these developments such as:

> ***Test differentiation from vaccination:*** People who received vaccination will test positive whether or not they are actually infected with *Borrelia burgdorferi*. These tests are aimed at distinguishing between these two situations.

Cytokine-based immunoassay: It could theoretically allow for earlier and more rapid diagnosis.

Rapid point-of-care using lateral flow technologies.

Metabolic biomarkers and biosignatures: Rather than wait for the antibodies to appear in the blood, investigators are examining a host of non-antibody molecules or biomarkers that arise earlier in infection. There appears to be a pattern of biomarkers that could differentiate early Lyme disease from the other illnesses tested, and could accurately diagnose the illness even in clinical samples that had tested negative by the current immunologic assay.

Other genome sequencing for multiple strains of Borrelia burgdorferi: Greater advances in diagnostics are anticipated as genetic information is combined with advances in microarray technology, imaging, metabolomics (a new science to identify and measure types and amounts of chemicals the body produces during illness), proteomics, next generation T-cell based measurements, and novel antigens.

Participate in clinical trials

Volunteering for a clinical trial is one way to help in the fight against LD. Studies need participants of different ages, sexes, races, and ethnicities to ensure that results are meaningful for many people.

To find out more about Lyme disease clinical trials and studies:

- Talk to your healthcare provider about local studies that may be right for you;

- Contact LD research centers clinics in your community;

- Search the ADEAR Center for clinical trials finder for a trial near you or to sign-up for email alerts about new trials;

- Sign up for a registry or matching service (such as TrialMatch) to be invited to participate in studies;

- Learn more about participating in Lyme disease research; and

- Watch a video about Lyme disease clinical trials.

As of the date of this writing (March 2020), there are 23 clinical trials on several aspects of disease and treatment of LD: 9 in the U.S. and 14 in Europe that are currently recruiting volunteers. For the reader's convenience, I have listed them below. (There are also 3 other trials that are not presently recruiting volunteers, so I have not listed them.) More information on any of these trials could be found in clinicaltrials.gov. Sidebar 19.1 presents as a sample the first of them in more details.

U.S. trials:

Novel diagnostics for early Lyme disease-causing
Last update post: 13 August 2019
Sponsors: Johns Hopkins University and MicroB-plex, Inc.

Xenodiagnosis after antibiotic treatment for Lyme disease
Last update post: 13 February 2020
Sponsors: (U.S.) National Institute for Allergy & Infectious Diseases (NIAID)

Evaluation, treatment, and follow-up of patients with Lyme disease
Last update post: 13 February 2020
Sponsor: (U.S.) National Institute for Allergy & Infectious Diseases (NIAID)

A comprehensive clinical, microbiological, and immunological assessment of patients with suspected Post-Treatment Lyme Disease Syndrome and selected control populations
Last update post: 18 January 2020
Sponsor: (U.S.) National Institute for Allergy & Infectious Diseases (NIAID)

Disulfiram: a test of symptom reduction among patients with previously treated Lyme disease
Last update post: 30 October 2019
Sponsor: Research Foundation for Mental Health, Inc.

Next generation sequencing of Lyme disease
Last update post: 16 October 2019
Sponsor: Stony Brooks University and Karlas, Inc.

Symptomatic management of Lyme arthritis
Last update post: 15 October 2019
Sponsor: Pittsburgh University

Rare disease patient registry and chronic fatigue at Stanford
Last update post: 6 Mars 2019
Sponsor: Stanford University

Myalgic encephalomyelitis chronic fatigue at the National Institutes of Health
Last update post: 18 January 2020
Sponsor: (U.S.) National Institute of Neurological Disorders & Stroke

European trials:

In Denmark:

Central nervous system infection in Denmark
Last update post: 6 May 2019
Sponsor: Aalborg University Hospital, Denmark

In France:

Peripheral facial paralysis sequelae of Lyme disease among children
Last update post: 12 June 2019
Sponsor: Centre Hospitalier Universitaire de Besancon, France

Understanding tick-borne diseases
Last update post: 15 August 2018
Sponsor: Institut Pasteur, France

Borrelia specie in cutaneous Lyme borreliosis
Last update post: 17 April 2019
Sponsor: University Hospital, Strasbourg, France

Lyme borreliosis and early cutaneous diagnostic
Last update post: 20 August 2019
Sponsor: University Hospital, Strasbourg, France

In Hungary:

Performance evaluation of the dualdur *in vitro* diagnostic system in the diagnosis of Lyme borreliosis
Last update post: 14 November 2019
Sponsors: Lyme Diagnostics Ltd and PharmaHungary Group, Hungary

In Slovenia:

Tick-borne encephalitis and positive borrelial antibodies
Last update post: 16 October 2018
Sponsor: University Medical Center, Ljubljana, Slovenia

Patient pre-treatment expectations about post-Lyme symptoms
Last update post: 15 May 2019
Sponsor: University Medical Center, Ljubljana, Slovenia

Inflammatory mediators in *erythema migrans*
Last update post: 5 May 2020
Sponsor: University Medical Center, Ljubljana, Slovenia

Different amoxicillin treatment regimens in *erythema migrans* patients
Last update post: 20 September 2020
Sponsor: University Medical Center, Ljubljana, Slovenia

Duration of doxycycline treatment in multiple *erythema migrans* patients
Last update post: 22 April 2019
Sponsor: University Medical Center, Ljubljana, Slovenia

Cytokines and chemokines in *erythema migrans*
Last update post: 22 April 2019
Sponsor: University Medical Center, Ljubljana, Slovenia

Study on early Lyme *neuroborreliosis*
Last update post: 16 October 2018
Sponsor: University Medical Center, Ljubljana, Slovenia

In Sweden:

Evaluation of cortisone treatment in children with acute facial nerve palsy
Last update post: 20 May 2019
Sponsors: Dalama County Council and Karolinska Institute, Sweden

The following are additional clinical trials listed by the American Lyme Disease Association (LDA). More information could be obtained from the LDA's website.

Rockefeller University Hospital, Center for Clinical and Translational Science
Last update post: 29 September 2019
Contact: The Rockefeller University Hospital Recruitment Office
Telephone: 212-327-7722
Email: RUCARES@Rockefeller.edu.

Columbia University, Lyme and Tick-Borne Diseases Research Center & Specimen Bank
Last update post: 20 April 2017
Contact: http://columbia-lyme.org/research/cr_reseach.html

New Columbia Lyme treatment study to reduce pain
Last update post: 6/23/2016
Contact: Ellen Brown at 646-774-8100 or eb3048@cumc.columbia.edu

Take-home points

- Becoming well-informed about the disease is one important long-term strategy. Also, programs that teach families about the various stages of Lyme and about ways to deal with caregiving challenges can help.

- Several private, professional, national, and international organizations (13 or more) offer invaluable advocating and supporting services.

- Begin treatment early in the disease process.

- Blood tests that check for antibodies (produced by the body to fight infection) to the bacterium that causes Lyme disease are not useful if done soon after a tick bite.

- According to the CDC&P, patients treated with appropriate antibiotics before the diagnostic tests are complete in the early stages of Lyme disease usually recover rapidly and completely.

- Look for new diagnostic tests.

- Participate in clinical trials. As of the date of this writing (March 2020), there are 26 open clinical trials recruiting volunteers for several facets of Lyme disease (9 in the U.S., 14 in European countries, and 3 through the Lyme Disease Association).

Sidebar 19.1

A sample clinical trial description: Novel Diagnostics for Early Lyme Disease

The following is the standard description provided by clinicaltrials.gov for the sample clinical trial below:

Novel Diagnostics for Early Lyme Disease (recruiting as of 13 August 2019)
Sponsor: MicroB-plex, Inc.
Collaborator: John Hopkins University

Brief Summary

There are more than 300,000 new cases of Lyme disease every year in the US. Lyme disease is a dangerous bacterial infection transmitted by tick bites and it becomes increasingly severe as the infection progresses. Definitive diagnosis is based on serum-based tests that have fundamental limitations: 1) current tests cannot detect early infections so patients do not receive antibiotic therapy until the infection has progressed, and 2) there is no way to measure if antibiotic therapy has been successful. MicroB-plex will address these two unmet clinical needs by introducing a novel, blood-based diagnostic method that will enable clinicians to diagnose infections earlier and to monitor the success of their interventions.

Detailed Description

Lyme disease

Lyme disease is the most commonly reported arthropod-borne infection in the U.S. with recent CDC&P estimates eclipsing 300,000 new cases in 2013. In addition to growing in frequency, the infections have a complex and increasingly severe course. Beginning with mild flu-like symptoms and frequently a signature bull's-eye rash (*erythema migrans*), Lyme disease can progress to severe articular, neurological, and cardiac symptoms, most of which are preventable with early antibiotic therapy. Leading investigators have identified two major shortcomings to the current serology-based methods for the definitive diagnosis of Early Localized Lyme disease. First, the clinical sensitivity in the first four weeks is poor, under 50% at the time of symptom onset, so many patients remain undiagnosed or unconfirmed until the disease has had time to progress. Second, serum antibody levels remain elevated long after the infection has been resolved making the monitoring of therapeutic success or diagnosis of re-infection virtually impossible. MicroB-plex will address these shortcomings by using a novel sample matrix from circulating antibody secreting cells (ASC) for diagnosis of Lyme disease. This novel matrix is MENSA (medium enriched for newly synthesized antibody). In this study, MicroB-plex and its clinical collaborators will test whether MENSA is effective in early Lyme diagnostic (within the first 2 weeks) and if this new approach will track therapeutic success.

Study design:
- Type: Observational (patient registry)
- Estimated enrollment: 100 participants
- Observational model: Case-control
- Time perspective: Prospective
- Target follow-up duration: 1 year
- Official title: Novel diagnostics for early Lyme disease
- Actual study start date: May 1, 2019
- Estimated primary completion date: December 31, 2020

Groups and cohorts:
- Group/cohort: Lyme infected
- Subjects: Presenting with suspected Lyme disease
- Controls: Subjects with no known Lyme disease, past or present

Intervention/treatment:
- Diagnostic test: Micro-Bplex Lyme immunoassay: Subject's blood and clinical data collected to develop a diagnostic immunoassay

Outcome Measures:
- Primary: Percentage of EM-positive patients who become MicroB-plex Lyme test positive prior to seroconversion
- Time frame: Within 14 days of enrollment, test measures anti-Lyme antibodies in MENSA becomes positive prior to conventional Lyme immunoassays that measure antibodies in serum, resulting in earlier diagnosis. Percentage of treated EM-positive patients who become MicroB-plex

Lyme disease

	Lyme test negative prior to their decline in serum, providing an earlier measure of effective therapy. Up to 1 year after enrollment.
Secondary:	Percentage of treated EM-positive patients who remain MicroB-plex Lyme test positive following treatment
Time frame:	Up to 1 year after enrollment, MicroB-plex test remains positive with treatment failure

Biospecimen Retention:

Samples With DNA anti-coagulated whole blood, serum, peripheral blood mononuclear cells.

Eligibility Criteria:

Choosing to participate in a study is an important personal decision. Talk with your doctor and family members or friends about deciding to join a study. To learn more about this study, you or your doctor may contact the study research staff using the contacts provided below.

Ages:	21-80 years (adult, older adult)
Sexes:	All
Accepts healthy volunteers:	Yes
Sampling method:	Non-probability sample

Study population:

Adult humans with a strong clinical suspicion of acute Lyme disease, with symptoms 7 days or less. Subjects will be recruited from medical centers residing in Maryland.

Inclusion criteria:

- Human adults at least 21 years of age and no more than 80 years of age at the time of screening.
- Have the ability to understand the requirements of the study, provide written informed consent (including consent for the use and disclosure of research-related health information), and comply with the study protocol procedures (including required study visits).
- High clinical suspicion of acute Lyme disease, with symptoms seven days or less.
- Must have *erythema migrans* rash and physician diagnosis of early Lyme disease.
- Brief history and physical exam will be obtained during the study visit.
- Be willing to return to our clinic for up to nine visits and blood draws over a one-year period.
- People who do not have Lyme disease, but want to participate as healthy controls (one-time visit)

Exclusion Criteria:

- Have poor venous access.
- Have received any immunosuppressive therapy including biologics or recent course of steroids,

or recent chemotherapy.
- Anti-TNF therapy (eg, adalimumab, etanercept, infliximab).
- Intravenous (IV) cyclophosphamide.
- Interleukin-1 receptor antagonist (anakinra).
- Intravenous immunoglobulin (IVIG).
- High dose prednisone or equivalent (> 100 mg/day).
- Plasmapheresis.
- Any new immunosuppressive/immunomodulatory agent.
- Any steroid injection (eg, intramuscular, intraarticular, or intravenous).
- On treatment for Lyme disease greater than seven days.
- Recent chemotherapy.
- History of solid organ transplant.
- History of autoimmune disorders (SLE, Rheumatoid arthritis, Scleroderma, etc.).
- History of inflammatory muscle disease (polymyositis, dermatomyositis, etc.).
- History of inflammatory bowel disease (Crohn's disease, Ulcerative colitis, etc.).
- History of HIV infection.
- Received a Lyme disease vaccine in the past.
- History of prior Lyme disease infection in the past.
- Have any condition that, in the opinion of the principal investigator, would significantly increase the risk for the subject.

Contacts and location:
Johns Hopkins University, School of Medicine, Baltimore, Maryland 21205, United States

Contact:
    Paul Auwaerter, MD, MBA
    Telephone: 443-287-4840
    Fishercenter@jhmc.edu

Principal Investigator:
    Paul Auwaerter, MD, MBA
    Telephone: 443-287-4840
    Fishercenter@jhmc.edu

Sponsors and collaborators: MicroB-plex, Inc.
    Principal Investigators:
    John L. Daiss, PhD
    Frances E. Lee, MD

Responsible party:
    MicroB-plex, Inc.

Lyme disease

ClinicalTrials.gov identifier: NCT039635

History of changes:
    Other study ID numbers: 12251305
    First posted: May 28, 2019
    Key recorded dates:
    Last update posted: August 13, 2019
    Last verified: August 2019

Individual participant data (IPD) Sgaring Statement:
    Plan to share IPD: No

Studies a U.S. FDA-regulated drug: No

Studies a U.S. FDA-regulated device: No

Product keywords provided by MicroB-plex:
    Lyme disease: Early diagnosis
    Borrelia burgdorferi: Treatments
    Antibody secreting: Outcome
    Cells: Human volunteers
    Diagnostics: Adults

Additional relevant MeSH terms:
    Lyme disease
    Borrelia burgdorferi
    Gram-negative bacterial infections
    Bacterial infections
    Tick-borne diseases
    Spirochaetales
    Infections

To learn more about this study, you or your doctor may contact the study research staff using the contact information provided by the sponsor. Please refer to this study by its ClinicalTrials.gov identifier (NCT number: **NCT0393635**).

20

Living with Lyme - The voice of the patient

I am briefly relating here several actual stories of persons (identified here as A, B, C,..) who had contracted Lyme disease and will follow with other serious situations of either being misdiagnosed or developing peripheral neuropathy. This latter situation may at times occur along with Lyme disease or, alternatively, be misdiagnosed for it.

Some familiar stories

The story of "A"

Two weeks after vacationing in a Lyme-prone area, "A" woke up with aches in the back, hips, legs, and stiff and creaky joints. In addition A had pain, fever, and felt very run down. Upon examination, A's doctor found the tell-tale bull's-eye rash around the belly button that A had not even noticed. A was diagnosed with LD, no blood tests were ordered, but a course of oral antibiotics was ordered. The fever went away and the general feeling was better, but the joint pain did not improve much and actually got worse over time with a lot of swelling in the knees and hips and the back still hurting. The doctor prescribed a second course of antibiotics, which is sometimes recommended when joint symptoms do not improve. The joint pain eventually improved, but the feeling of tiredness and lethargy remained for several weeks to months. Knowing how long recovery may take, one should be more careful than ever about preventing tick bites and checking one's body every day.

The story of "B"

During a family camping trip, "B" got LD but never saw a tick or a rash and did not have symptoms until suddenly experiencing very serious knee and joint pains. While walking up a flight of stairs, B suddenly began to experience pain in both knees and stiffness throughout the entire body, knees, ankles, wrists, and shoulders. Just walking within the house was a painful ordeal. B had vacationed the month before in a camping area where ticks were crawling but did not recall getting the bull's eye rash of LD. B's doctor ordered a blood test that came back positive for LD and further ordered an antibiotic (doxycycline) that, within a week's time, resolved B's symptoms. For the next camping trip, B then decided to use repellent on the body and wear permethrin-treated clothing. Importantly, B further decided not to delay consulting the doctor if the same signs and symptoms of LD should recur.

The story of "C"

"C" got LD twice! The first time, C developed a bull's eye-rash on the back of the knee followed by one on the lower back a few days later, but never found a tick. A search on the internet and the CDC&P's website convinced C to consult a doctor. A blood test was ordered. Five days before the blood test came back, C rapidly developed flu-like symptoms with a high fever, a horrible headache, and muscles hurting. After a course of antibiotics, the symptoms cleared up very quickly within a couple of days. The test did eventually come back positive. About 3-4 years ago, C had gotten LD in the form of a scab on the hip that would not go away. A small rash had developed around the scab area. C's doctor prescribed a course of antibiotics even before the blood test results came back positive. Having taken the antibiotics right away, C never got sick like the first time and never had any other symptoms. C now insists on using a tick-repellent and self-examining for the presence of ticks or a rash.

The morale of the stories of A, B, and C is to get accurate information about the signs of LD. If bitten by a tick or having a suspicious rash, consult a doctor the sooner the better to be diagnosed. If LD is suspected, get a course of antibiotics even before blood test results confirm the diagnosis, thus avoiding the likelihood of developing more serious symptoms.

Other serious situations

Feeling worse after treatment? May be it is not LD!

"D" received years of treatment for "chronic Lyme disease (CLD)". Over time, the symptoms worsened and D began to lose sight in the right eye. Persistent symptoms may indicate that it may not be LD after all and, indeed, D was diagnosed with a pituitary tumor. Due to the increase in growth hormone levels during the three years of improper treatment for LD, D had permanent musculoskeletal changes and suffers from heart disease, joint pain, and severe obstructive sleep apnea.

The morale of the story of D is, if feeling worse after treatment, consult or be referred to an LD specialist for it may not be LD.

The following cases deal with neuropathy associated with the peripheral nervous system.

What is PN-associated with PNS?

Peripheral neuropathy (PN) results from damage to the peripheral nervous system (PNS). It commonly affects nerves located within the extremities such as toes, feet, legs, fingers, hands, and arms. More than 100 types have been identified. The resulting sensory nerve damage is common and difficult-to-manage. It may lead to an over-sensitization of nerves causing severe pain from stimuli that are normally painless. It may follow different patterns which vary over a period of days, weeks, or years.

There is no cure for neuropathy. It affects an estimated 20 million people in the U.S. Most people who develop PN are over age 55, but people can be affected at any age.

PN pain characterization

The pain of PN-associated PNS is: sudden, sharp, stabbing, throbbing, burning, intense, constant, continuous or chronic, debilitating, concentration-stealing, unpredictable, extremely sensitive to touch, variable, life-changing...and other adjectives depending on the affected patient.

Common neuropathic symptoms are:
- Numbness, tingling, electric shocks, pins and needles, muscle spasms, temperature sensitivity, stabbing, and stinging sensations.
- Pain: in the legs, arms, feet, and hands.
- Pain progression: from "episodes of burning legs" to "burning in the face and jaw" to "painful stabbing feelings" in one side of the chest and the arm, mouth and facial pain due to burning mouth syndrome (BMS - a rare disease associated with peripheral neuropathy and trigerminal neuralgia), buttocks, groin, and neck.
- Balance issues: due to numbness and lack of awareness of one's feet and legs while standing, and lack of ability to focus and concentrate. Sleeping difficulties, daytime sleepiness, and constant fatigue.

PN symptoms variability

PN symptoms vary greatly depending upon the type of nerves that have been damaged. This condition may cause: Numbness, Prickling, or Tingling sensations (acronym **NPT**).

PN treatment

Neuropathy is typically treated by addressing the causes and working on symptom relief. Treatment is often a matter of trial-and-error. A treatment that may work well for one person may be totally ineffective for another. Persistence and perseverance may be the key contributors to successfully finding help. A key place to start in treating neuropathy is to limit the effects of the cause, if the cause is known. If a medication or toxin is involved, limiting or eliminating exposure will help. Sometimes, addressing the underlying cause may be all that is needed to reverse or halt the spread of symptoms. It is important to always work with your doctor in developing a treatment plan.

MEDICATIONS:

Prescription medications for the treatment of neuropathy typically fall into two categories: antiseizure medications and antidepressants. Such medications work by affecting the electrical and chemical activity of the nerves. Unfortunately, most pain and nerve medications can have side effects of drowsiness and fogginess. Some people have had great success with the use of medications made by compounding pharmacists, typically medication in an ointment form that is applied to the affected area. Others who experience side effects from an oral medication may tolerate it better in a compounded application where the medication stays local and immediately enters the bloodstream by bypassing the digestive system.

Prescriptions and over-the-counter drugs vary widely in effectiveness, often leading to limited or decreased benefits over time, have side effects even when the treatment is effective, and may require balancing the need to increase medication dosage and frequency.

Many patients use customized treatment regimens aimed at managing symptoms that impact them the most. These include: medications, special procedures, specific devices, non-drug therapies, and lifestyle changes. There may be insurance restrictions on some medications. The *"ideal pain medication would be non-sedating, not habit-forming or addictive, tolerated for long-periods of time without needing increases in dosage, affordable, and have minimal health impacts when used long-term...."* a tall order not yet achieved! I briefly review each of these treatment approaches below. A more detailed review would take us very far from our objective.

- *Antidepressants*; Cymbalta (duloxetine), Doxepin (sinequan), Effexor (venlafaxine), Elavil (amitriptyline), and Zoloft (sertraline). They offer varying levels of pain control ranging from "modestly effective" to "absolutely worthless".

- *Anticonvulsants:* Keppra (leviracetam), Lyrica (pregabalin - recent prescription medication specifically approved for the treatment of neuropathy pain), Neurontin (gabapentin - widely used with great effect in some), Tegretol (carbamazepine), Topamax (topiramate), and Vimpat (Lacosamide) provide tolerable pain management but do not completely eliminate the pain. For some, a modest dose may allow a level of functionality not otherwise possible. Others may become "maxed out of approved levels with each medication" as there is a need to increase dosages regularly to obtain pain relief. There are difficulties with withdrawal symptoms following attempts to reduce medication dosage. There is also concern over the long-term side effects of anticonvulsants' use. However, there are certain downsides including "brain fog" (a variety of cognitive effects such as drowsiness, difficulty focusing, and short-term memory loss –- see Sidebar 3.4), dizziness, fatigue, and weight gain.

- *Pain relievers:* Are frequently used in addition to PN medications. They include:

  - *Opioids:* Cannabinoid oil, hydrocodone, lidocaine (nightly patches), medical marijuana, methadone, morphine, oxycodone, tramadol (when burning or stabbing pain sensations become too intense),

  - *Opioid combination products:* Percocet (oxycodone and acetaminophen) and Vicodin (hydrocodone and acetaminophen);

  - *Other prescription drugs:* diclofenac, fentanyl, tramadol; and

  - *Over-the-counter pain medications:* non-steroidal anti-inflammatory drugs (NSAIDs) such as acetaminophen (Tylenol), Ibuprofen, Naproxen (Aleve) may provide relief to some. In addition, various creams and ointments have proved helpful or soothing.

- ***Others:*** Muscle relaxants, IVIg (intravenous immunoglobulin); sleep aids (Benzodiazepines); steroids; and compounded pain relief creams.

THERAPIES:

Various therapies have proven helpful to some:

- ***Epidurals;***
- ***Laser treatments;***
- ***Plasmapheresis;*** and
- ***Stem cell therapy.***

SPECIFIC DEVICES:

- ***Acupuncture;***
- ***Cervical traction:*** In addition to epidurals (only for temporary alternatives to pain reduction);
- ***Ice, heat, hydrotherapy;***
- ***Inversion table;***
- ***Low- level light therapy***;
- ***Orthopedic footwear;***
- ***Physical therapy;***
- ***Rebuilder;***
- ***TENS*** (Transcutaneous Electrical Nerve Stimulation); and
- ***Therapeutic beds and pillows.***

In addition to medical treatments, non-medical therapies could also be beneficial such as:

NON-DRUG THERAPIES

Non-drug therapies include alternative and complementary medicine, Epsom salt, homeopathic remedies, massage (including cranio-sacral therapy and reflexology), meditation.

LIFE-STYLE MODIFICATIONS:

These include diet, exercise, meditation, and mindfulness.

SUPPLEMENTS:

Many supplements (herbal and others) have been touted as helpful for different conditions in recent years. Alpha lipoic acid has been used for years in Europe as a treatment for neuropathy. For those with neuropathy caused or exacerbated by nutritional deficiencies, supplements can be vital.

PAIN MANAGEMENT CLINICS:

Medical groups, hospitals, and independent organizations may offer pain management clinics. These facilities include specialized staff that can help patients try various types of medications and therapies in order to find treatments that work best for them.

Patients' daily experiences and struggles

These include but are not limited to:

- *Numerous pain sensations:* most significant symptoms are numbness (with or without pain, continuous or not), tingling (with or without the sensation of an electrical shock), burning sensation, and stabbing in various locations of the body (particularly the hands, feet, neck, face, shoulders, and stomach) with a variation in the frequency, intensity, and type of pain ("sharp", "constant", "bruising", "crushing", "soreness", and "flares") with interrelation between various pain sensations. Pain may be continuous (24 hours a day, 7 days a week) or sporadic, may be different moment-to-moment, or else more of a week-to-week or a month-to-month fluctuation, and may depend upon the level of activity throughout the day or the position (whether up or sitting) and level of tiredness.

- *Other symptoms:* issues with balance and coordination due to the lack of awareness of certain areas of the body and the surroundings; inconsistent sleep patterns or/and insomnia; hot or cold sensitivity; prickling sensations; soreness; itching; cramps; inability to perform activities even when sitting; and driving difficulties.

- *Sudden and unpredictable neuropathic pain:* limits the ability to obtain adequate pain control on a consistent basis; and

- *Difficulty in achieving pain relief:* even using both drug and non-drug treatments to control symptoms with a vast range of variability in effectiveness. (There is a significant burden of trial-and-error therapy regimens, difficulty of weighing benefits and adverse effects when making treatment decisions, and challenges faced in obtaining access to prescription drug products).

Patients' needs

There are several needs, including:

- *Increased awareness and research of neuropathic pain:* associated with peripheral neuropathy:

- *Concerns and frustration regarding the lack of understanding in the medical community:* this is to be added to the difficulties of finding doctors who understand their condition, being prescribed helpful medications, and the stigma surrounding the use of medications that target pain.

- *"Broad" pain control:* this should cover a wide spectrum of painful sensations and reduce the amount of drug products needed to cover the full range of painful neuropathy symptoms while reducing the number and severity of long-term and short-term side effects. Such an "ideal" treatment should incorporate (1) drug products that target symptomatic pain (e.g., reducing sensations of pain) and (2) disease-modifying drug products that slow or reverse the loss of feeling and numbness.

Adverse impacts on patients' daily lives

NP has a debilitating impact on patients' and their families' daily lives:

- *Inability to care:* for oneself, one's children, managing the household, partaking in daily activities.

- *Mood changes:* becoming very intolerant, impatient, and short-tempered when interacting with others.

- *Emotional burdens:* severe periods of emotional strain, periods of frequent crying, suicidal thoughts, extreme hopelessness, and depression. Being stigmatized as a "drug seeker".

- *Loss of family connections:* decreased quality time with family, feelings of hopelessness, and reduced sexual intimacy within the marriage. and close relationships.

- *Limited social relations:* limited social interactions due to lack of understanding of peripheral neuropathy by those around the patients, lack of social engagements, isolation (fears of remaining mainly "house-bound" or "bed-ridden"),

- *Work and career:* lagging job performance due to several inabilities to work at all to limitations on time at work, walk and stand for extended periods of time, drive, cognitive issues, lack of concentration, and lack of energy.

- ***Worry about the future:*** particularly regarding disease progression and the consequent inability to adequately manage the pain as the symptoms worsen, reduction in the ability to perform activities.

- ***Current treatment challenges:*** limited remaining treatment options and challenges of managing the many symptoms while making difficult decisions about drug and non-drug treatments.

Take-home points

- If bitten by a tick or having a suspicious rash, consult a doctor the sooner the better to be diagnosed. If Lyme disease is suspected, get a course of antibiotics even before blood test results confirm the disease.

- Peripheral neuropathy results from damage to the peripheral nervous system. It commonly affects nerves located within the extremities. More than 100 types have been identified. The resulting sensory nerve damage is common and difficult-to-manage. It may lead to an over-sensitization of nerves, causing severe pain from stimuli that are normally painless. It may follow different patterns which vary over a period of days, weeks, or years.

- Peripheral neuropathy affects an estimated 20 million people in the U.S., affecting mostly people over age 55 but also others of any age.

- The pain of peripheral neuropathy associated with the peripheral nervous system has been described as: sudden, sharp, stabbing, throbbing, burning, intense, constant, continuous or chronic, debilitating, concentration-stealing, unpredictable, extreme sensitivity to touch, variable, and life-changing.

- Common neuropathic symptoms are progressive and include: numbness, tingling, electric shocks, pins and needles, muscle spasms, temperature sensitivity, stabbing and stinging sensations, pain in the extremities, balance issues, sleeping difficulties, daytime sleepiness, and constant fatigue.

- Treatments include medications, special procedures, specific devices, non-drug therapies, and lifestyle changes. The "ideal pain medication would be non-sedating, not habit- forming or addictive, tolerated for long-periods of time without needing increases in dosage, affordable, and have minimal health impacts when used long- term". Note that there may be insurance restrictions on some medications as to payments or reimbursements.

- Many patients use customized treatment regimens aimed at managing symptoms that impact them.

- Patients' daily experiences and struggles include but are not limited to numerous pain sensations sudden and unpredictable neuropathic pain, difficulty in achieving pain relief. Even using both drug and non-drug treatments to control symptoms, there can be a vast range of variability in effectiveness.

- Patients' needs include increased awareness and research of neuropathic pain associated with peripheral neuropathy, "broad" pain control to cover a wide spectrum of painful sensations and reduce the amount of drug products needed to cover the full range of painful neuropathy symptoms while reducing the number and severity of long-term and short-term side effects.

- The "ideal" treatment should incorporate drug products that target symptomatic pain (e.g., reducing sensations of pain) and disease-modifying drug products that slow or reverse the loss of feeling and numbness.

- Patients confront several adverse impacts in their and their families' daily lives including the inability to care for oneself and one's immediate family, mood changes, emotional burdens, loss of family connections, limited social relations, work and career, worry about the future, and current treatment challenges among others.

21

Any recent out-of-the-box research developments?

I report here on the recent most interesting out-of-the-box research developments.

IN LYME DISEASE

Repeated Lyme infection has an immune priming effect

In a multi-centered randomized clinical trial, a research team led by Dr. B.A. Fallon and his team at Columbia University, New York examined a cohort of 179 participants including (a) 24 with recent *erythema migrans* rash but without prior Lyme disease, (b) 119 with persistent post-treatment Lyme symptoms, and (c) 28 seronegative (negative blood tests) controls with no prior Lyme history. They found that anti-neuronal autoantibodies were elevated in those with prior Lyme history but not in those without prior Lyme history, suggesting an immune priming effect of repeated infection. Future prospective studies will determine whether these autoantibodies emerge after Lyme infection and whether their emergence coincides with persistent neurologic or neuropsychiatric symptoms.

A new cognitive fingerprint has been identified for post-treatment Lyme disease

In a multi-centered randomized clinical trial, Dr. J. Kalp of Columbia Lyme Disease Research Center, Columbia University, New York found that the cognitive profile of patients with post treatment Lyme disease is meaningfully different from the profile of patients with major depression. Although both groups might have fatigue and mental fogginess, the Lyme group more often reports problems with verbal memory and verbal fluency while the depressed (non-Lyme) group more often reports slower processing speed and poor attention. These results highlight the value of neurocognitive testing in helping to tease out the potential causes of cognitive problems in patients with post-treatment Lyme disease.

Lyme can persist in the human body even after extensive antibiotic treatment

In a multi-centered analysis of human autopsied tissues (of the brain, heart, kidney, and liver) of a deceased patient after a 16-year long illness with Lyme, using several histological and immunohistochemical methods (confocal microscopy, fluorescent *in situ* hybridization-FISH, polymerase chain reaction-PCR, and whole genome sequencing-WGS/metagenomics), Sapi I *et al.* of the University of New Haven, Connecticut found that *Borrelia burgdorferi* can persist in the human body, not only in the spirochetal but also in the antibiotic-resistant biofilm form, even after long-term antibiotic treatment, and the organism in biofilm form might further trigger chronic inflammation.

Vancomycin is an effective antibiotic for killing *Borrelia burgdorferi*

In post-treatment Lyme disease, the persistence of symptoms after antibiotic treatment has never been conclusively explained. It is possible that human *Borrelia burgdorferi* infection results in residual, persistent infection even after treatment due to the continued presence of bacteria and/or an ongoing immune response. To identify antibiotics that might completely eradicate bacterial replication, Lewis K and his team followed up earlier observations that beta-lactam antibiotics might be effective. They found that vancomycin is effective against *in vitro* cultures of growing *Borrelia burgdorferi,* cell wall synthesis still occurred, and could be blocked by vancomycin. *In vivo,* vancomycin and ceftriaxone each completely blocked bacterial growth, compared with partial eradication by doxycycline. These studies suggest that more effective antimicrobial drugs, used early in infection, may prevent or reduce the occurrence of persisting infection.

Drug-tolerant persister cells in Lyme can be eradicated by an appropriate combination of antibiotics

When treated early, Lyme disease usually resolves, but when left untreated, it can result in symptoms such as arthritis and encephalopathy, requiring multiple courses of antibiotic therapy. Given that antibiotic resistance has not been observed for *Borrelia burgdorferi*, the reason for the recalcitrance of late-stage disease to antibiotics is unclear. Drawing on the fact that in other chronic infections, drug-tolerant "persister" cells (*borrelia* organisms remaining despite antibiotic therapy), are associated with disease-recalcitrance, a research team led by Dr. B. Sharma investigated the presence of, and found such persisters in Lyme disease. The combination of daptomycin (a membrane-active bactericidal antibiotic), mitomycin C (an anticancer agent), and ceftriaxone killed all persisters and eradicated all live bacteria. An independent research team led by Dr. J. Feng confirmed this result with a different combination of antibiotics (daptomycin, cefoperazone, and doxycycline). These findings may have implications for improved treatment of Lyme disease. However, further studies are needed to validate whether such combination antimicrobial approaches are useful in human infection.

Rodent-targeted Lyme vaccine shows decrease in Lyme disease transmission

A field trial study by the Connecticut Agricultural Experiment Station (CAES) and US Biologic, Inc. showed the effectiveness of an orally-delivered anti-Lyme vaccine that targets the white-footed mouse, the major wildlife source of Lyme disease. The study took place in the residential area of Redding, Connecticut, over a three-year time period. It showed substantial decreases in the number of infected mice. One year into the study, test sites that had been treated with the vaccine showed a 13X greater decrease in black-legged ticks (*Ixodes scapularis*, the primary vector associated with the spread of disease) infected with *Borrelia burgdorferi* (the bacterium that causes Lyme disease) compared to control sites (i.e., 26% drop versus 2% drop).

A second effect showed that the vaccine causes the mice to generate antibodies so that previously infected ticks act as a 'xenodiagnostic marker' of vaccine impact, meaning once they ingest the antibodies, while feeding on vaccinated mice, the ticks become 'cleared' of infection. Non-infected mice fed on vaccine-coated pellets are then protected from the *Borrelia burgdorferi* infection and cannot pass the disease to other animals and humans.

Study shows increase of non-Lyme tick-borne diseases

Dr. Lee-Lewandowski and her team evaluated trends in non-Lyme tick-borne disease (NLTBD). Her study lasted over the course of seven years (2010-2016), including polymerase chain reaction (PCR) and serological tests. It showed that: (a) testing and positivity for most NLTBDs increased dramatically over the course of the study, (b) the number of positive cases generally exceeded those reported by the CDC&P, (c) the frequency of NLTBD in the U.S. is seasonal, but testing activity and positive test results are present throughout all months of the year, and (d) positive results for NLTBD testing originated primarily from a limited number of States, signifying a geographic concentration and distribution. Although the findings are not surprising, their increased awareness will result in the consideration of other tick-borne diseases by the medical profession.

IN LYME-ASSOCIATED PERIPHERAL NEUROPATHY

Medical marijuana as a second or third treatment line for peripheral neuropathy

Marijuana refers to strains of the cannabis plant that contain substances, called cannabinoids, that can have effects on the human body. Medical marijuana refers broadly to marijuana products that are used for medicinal purposes. The two main cannabinoids are: Tetrahydrocannabinol (THC -the psychoactive compound) and Cannabidiol (CBD-no psychoactive effect). THC and/or CBD binds to specific cannabinoid receptors in the brain and peripheral nervous system, which is likely the way in which it is able to alleviate chronic pain. Because medical marijuana products vary widely, no consistent recommendations can be made on dosing although the recommendation is usually "*start low, go slow, and stay low.*" One study found that 25 mg herbal cannabis with 9.4% THC, administered as a single smoked inhalation three times daily for five days, significantly reduced average pain intensity compared with a 0% THC cannabis placebo in adult participants with chronic post-traumatic or post-surgical neuropathic pain.

Medical marijuana is most often allowed to be used in debilitating conditions such as neuropathy, spinal cord injury with spasticity, multiple sclerosis, epilepsy, amyotrophic lateral sclerosis, chronic pain, Parkinson's disease, Huntington's disease, HIV/AIDS, and cancer; and for controlling symptoms and conditions such as muscle spasms, severe nausea, cachexia, post-traumatic stress disorder, etc. However, there are currently only three studies that have shown significant efficacy in 3 main medical conditions (1) chronic pain, including neuropathic pain, (2) chemotherapy-induced nausea and vomiting, and (3) multiple sclerosis-related spasticity. One of the neuropathic pain conditions that has been evaluated by randomized, double-blind studies, is HIV-neuropathy, in which studies have shown promising effects on pain.

Unlike opioid pain medicines, medical marijuana is not lethal in overdose. Further, unlike tobacco, smoking cannabis does not increase the risk for certain cancers (i.e., lung, head, and neck) in adults.

However, it is associated with chronic cough. Other side effects of particular relevance in peripheral neuropathic pain are: impairment in learning, attention, and memory; developing or worsening substance use disorders (including alcohol, tobacco, and other illicit drugs); psychiatric effects, such as development of social anxiety disorder, increased risk of suicidal thoughts, increased risk of schizophrenia, and other psychoses, and worsening of pre-existing bipolar symptoms with daily use.

In the treatment of chronic neuropathic pain, there is some efficacy of medical marijuana/cannabis. Clinical trials of different routes of administration (sublingual, oral, smoked, and vaporized) have demonstrated analgesic benefit. Some treatment guidelines for neuropathic pain recommend consideration of cannabinoids as a second- or third-line agent.

As expenses for medical marijuana are not covered by health plans and are out-of-pocket, knowing the cost-effectiveness of medical marijuana may impact patients' decisions regarding its use. Inhaled cannabis appears to be cost-effective when used as second- or third-line treatment.

Viral gene therapy blocks peripheral nerve damage in mice

Nerve axons serve as the wiring of the nervous system, sending electrical signals to control movement and sense of touch. When axons are damaged, whether by injury or as a side effect of certain drugs, a program is triggered that leads axons to self-destruct. This destruction likely plays an important role in multiple neurodegenerative conditions, including peripheral neuropathy, Parkinson's disease, and amyotrophic lateral sclerosis (ALS). For the first time, using a mutated form of the protein SARM1, scientists at the Washington University School of Medicine in St. Louis have developed a standard viral gene therapy that blocks this process, preventing axon destruction in mice and suggesting a therapeutic strategy to help prevent peripheral neuropathy. However, this treatment needs to be tested in human clinical trials.

A similar viral gene therapy is now in clinical trials for a genetic disorder called Duchenne muscular dystrophy (DMD) using a different protein delivered to address muscle loss, but the virus is the same. In theory, it could be possible to change the viral packaging to direct the viruses to deliver their gene payload to different types of cells — sensory neurons for peripheral neuropathy or motor neurons for amyotrophic lateral sclerosis, for example.

Stem cell treatment is still a hope in neuropathy pain treatment

Stem cells might theoretically be applied to any number of medical conditions by growing them in a way that repairs or replaces specific types of damaged or diseased human tissue but we are still far from being able to cure illnesses using them. However, there are several ongoing clinical trials and research is extensive. My recommendation is to proceed with caution. If a patient finds a reputable medical research institution or a clinic affiliated with it that is looking at stem cell treatments for peripheral neuropathy, it could be a good opportunity. One must bear in mind, however, that successful treatment is not guaranteed and may be accompanied with risks or/and side effects. Proceeding otherwise may be extremely expensive and perhaps carry health risks.

Blocking the pathway of neuropathy pain may lead to a new treatment approach

In an international collaboration between Indiana University in Bloomington and the Turku Center for Biotechnology in Finland, researchers identified the correlation between the formation of a certain protein in the cell (NOS1AP) and the transmission of neurological pain. By disrupting the formation of this protein and the resultant biological pathway, pain is reduced. This treatment may be feasible without the severe side effects attributable to current drugs. However, the formation of the NOSiAP protein and its contribution to chronic pain warrants more study in the quest for chronic pain relief.

Scrambler therapy for treating neuropathic pain

Scrambler therapy (marketed as Calmare™ therapy in the U.S.) is a new type of FDA-approved, non-toxic pain relief that uses a rapidly changing electrical impulse to send a "non-pain" signal along the same pain fibers that are sending the "pain" stimulus. No adverse effects have been noted so far but substantial pain relief was reported (50% reduction in low back pain, 91% in pain from failed back syndrome, post herpetic neuropathy, and spinal cord stenosis). The procedure is expensive (~ $100,000 per treatment series), unless it is part of a clinical trial, and not covered by insurance.

Neuropathy pain medication could end the opioid crisis

As opioid-related deaths and addiction in the U.S. reach epidemic proportions, a new kind of neuropathy pain medication is being developed that would not have the devastating side effects often caused by commonly prescribed drugs (such as Oxycontin). The pursuit is difficult because the very mechanisms that make those pills good at dulling pain are the ones that too often lead to crippling addiction and drug abuse. Like their close chemical cousin heroin, prescription opioids can cause people to become physically dependent on them (that is, "separating the addictive properties of opiates from the pain-reducing properties").

Flu shots will not make neuropathic pain worse

According to the CDC&P, there is no evidence that the flu shot will make neuropathy symptoms worse unless the person had a vaccine-induced neuropathy associated with Guillain-Barre Syndrome (GBS), a rare disorder caused by damage of the peripheral nerves featuring weakness, paralysis, and sometimes even breathing issues. Infections commonly precede GBS and are thought to "trigger" it through a process called "molecular mimicry", an autoimmune disease. The flu vaccine works by stimulating the immune system to produce antibodies that actually fight the virus, it does not give the flu. In fact, CDC&P strongly recommends that peripheral neuropathy patients (excluding GBS) receive a flu shot every year because they are more prone to developing serious complications if they get the flu.

Peripheral neuropathy is associated with mitochondrial and bioenergetic dysfunctions

Researchers at McGill University in Montreal, Canada recently revealed that peripheral neuropathy (nerve damage) is associated with mitochondrial and bioenergetic impairments. Mitochondria are small cellular organelles in the body responsible for the production of energy through the process of respiration. Mitochondria dysfunction can impair the cellular energy metabolism, affecting several organs and cells, like neurons, potentially leading to cell death and neurodegeneration. The lack of energy can cause a reduction in the activity of sodium-potassium pumps that ultimately lead to a condition that triggers the spontaneous activity of sensory neurons, resulting in neuropathic pain. The research team suggested that therapeutic agents able to normalize these two dysfunctions might have a clinical benefit in patients experiencing pain due to injured nerves.

IN LYME ENCEPHALOPATHY

Intravenous ceftriaxone therapy improves short-term cognition for patients with post-treatment Lyme encephalopathy

In a clinical trial led by Dr. B. A. Fallon from Columbia University in New York, enrolled patients had mild-to-moderate cognitive impairment and marked levels of fatigue, pain, and impaired physical functioning. It was found that IV ceftriaxone therapy provides short-term cognitive improvement for patients with post-treatment Lyme encephalopathy, but relapse in cognition occurs after antibiotic is discontinued. Treatment strategies that result in sustained cognitive improvement are needed.

IN LYME CARDITIS AND ARTHRITIS

***Borrelia* may be autoimmunity-mediated in Lyme carditis and arthritis**

A research team led by Dr. E. S. Raveche of the University of Medicine & Dentistry in Newark, New Jersey showed that certain manifestations of Lyme disease such as arthritis and carditis may be autoimmunity-mediated due to the molecular mimicry between the bacterium *Borrelia burgdorferi* and its self-components. There is immunological cross-reactivity, suggesting that *Borrelia burgdorferi* may share common epitopes which mimic self-proteins. These implications could be important for certain autoimmunity-susceptible individuals or animals that become infected with *Borrelia burgdorferi*.

IN CHILDREN AND ADOLESCENTS

***Borrelia burgdorferi* persists in the gastrointestinal tract of children and adolescents with Lyme disease**

Dr. M Fried and his team at Jersey Shore Medical Center in Neptune, New Jersey documented the persistence of *Borrelia burgdorferi* DNA in the gastrointestinal tract of pediatric patients who have already been treated with antibiotics for Lyme disease even after antibiotic treatment.

Children with chronic Lyme disease have cognitive deficits

Although neurologic Lyme disease is known to cause cognitive dysfunction in adults, little is known about its long-term sequelae in children. In a controlled study, Dr. F. A. Tager and her team at Columbia University, New York found that children with Lyme disease have significantly more cognitive and psychiatric disturbances. Cognitive deficits were also found after controlling for anxiety, depression, and fatigue. Lyme disease in children may be accompanied by long-term neuropsychiatric disturbances, resulting in psychosocial and academic impairments.

Take home points

- Repeated Lyme infection has an immune priming effect. However, prospective studies will determine whether these autoantibodies emerge after Lyme infection and whether their emergence coincides with persistent neurologic or neuropsychiatric symptoms.

- A new cognitive fingerprint has been identified for post-treatment Lyme disease, highlighting the value of neurocognitive testing in helping to tease out the potential causes of cognitive problems in patients with post-treatment Lyme disease.

- Lyme can persist in the human body not only in the spirochetal but also in the antibiotic-resistant biofilm form, even after long-term antibiotic treatment. In addition, the organism in biofilm form might further trigger chronic inflammation.

- *Borrelia burgdorferi* infection results in residual, persistent infection even after treatment due to the continued presence of bacteria and/or an ongoing immune response. Vancomycin is an effective antibiotic for killing *Borrelia burgdorferi*. More effective antimicrobial drugs, used early in the infection, may prevent or reduce the occurrence of persisting infection.

- When treated early, Lyme disease usually resolves, but when left untreated, it can result in symptoms such as arthritis and encephalopathy, requiring multiple courses of antibiotic therapy. Drug-tolerant persister cells in Lyme can be eradicated by an appropriate combination of antibiotics.

- The combination of daptomycin (a membrane-active bactericidal antibiotic), mitomycin C (an anticancer agent), and ceftriaxone can kill all persister cells and eradicate all live bacteria. However, further studies are needed to validate whether such combined antimicrobial approaches are useful in human infection.

- Medical marijuana is most often used in debilitating conditions such as neuropathy and has merit as a second or third treatment line for peripheral neuropathy. Unlike opioid pain medicines, medical marijuana is not lethal in overdose. Further, unlike tobacco, smoking cannabis does not increase the risk for certain cancers. However, it is associated with chronic cough and other side effects (impairment in learning, attention, and memory; development or worsening of substance use disorders, psychiatric effects, suicidal thoughts, increased risk of schizophrenia, and other psychoses, and worsening of pre-existing bipolar symptoms with daily use. However, expenses for medical marijuana are not covered by health plans and are out-of-pocket. Inhaled cannabis appears to be cost-effective when used as second- or third-line treatment.

- Viral gene therapy blocks peripheral nerve damage in mice by preventing axon destruction. However, this treatment needs to be tested in human clinical trials.

- Stem cells might theoretically be applied to any number of medical conditions by growing them in a way that repairs or replaces specific types of damaged or diseased human tissue, but we are still far from being able to cure illnesses using them. Stem cell treatment is still a hope in neuropathy pain treatment when conducted in a reputable medical research institution. However, successful treatment is not guaranteed and may be accompanied with risks or/and side effects.

- By disrupting the formation of a certain protein in the cell (NOS1AP) and blocking the resultant biological pathway, pain neuropathy is reduced, leading to a new treatment approach

- Scrambler therapy is a new type of FDA-approved, non-toxic pain relief that uses a rapidly changing electrical impulse to send a "non-pain" signal along the same pain fibers that are sending the "pain" stimulus. No adverse effects have been noted so far but substantial pain relief was reported. The procedure is expensive and not insurance-covered.

- As opioid-related deaths and addiction in the U.S. reach epidemic proportions, a new kind of neuropathy pain medication is being developed that would not have the devastating side effects often caused by commonly prescribed drugs (such as oxycontin). Neuropathic pain medication could end the opioid crisis.

- According to the CDC&P, there is no evidence that the flu shot will make neuropathy symptoms worse unless the person had a vaccine-induced neuropathy associated with Guillain-Barre Syndrome.

- Peripheral neuropathy is associated with mitochondrial and bioenergetic dysfunctions. Therapeutic agents able to normalize these two dysfunctions might have a clinical benefit in patients experiencing pain due to injured nerves.

- Intravenous ceftriaxone therapy improves short-term cognition for patients with post-treatment Lyme encephalopathy, but relapse in cognition occurs after antibiotic is discontinued.

- *Borrelia burgdorferi* may be autoimmunity-mediated in Lyme carditis and arthritis, making it important for certain autoimmunity-susceptible individuals or animals that become infected with it.

- *Borrelia burgdorferi* persists in the gastrointestinal tract of children and adolescents with Lyme disease even after antibiotic treatment.

- Children with chronic Lyme disease have cognitive and psychiatric disturbances that may result in psychosocial and academic impairments.

22
Latest news in the world of Lyme

(U.S.) Government actions

President signs Tick Act into law

On December 20, 2019, President Trump signed the Kay Hagan Tick Act into law. As a result, Lyme will get some monies for the development of a vector-borne disease national strategy and also for CDC&P's grants to States to help provide funding for surveillance and other vector-borne diseases issues. The bill includes all vector-borne diseases, mosquito, and tick-borne.

President's FY'21 budget includes increase for vector-borne diseases

The White House released the Fiscal Year 2021 Budget proposal which includes $66 million for the CDC&P's vector- borne disease activities (a $14 million increase compared to the 2020 enacted level) that focuses on tick-borne diseases. The Budget also invests in the National Institutes of Health (NIH) research to improve the Nation's understanding of vector-borne diseases. Excerpts:

- *"Prioritizes Critical Health Research and Supports Innovation"* … "NIH would continue to address the opioid epidemic and emerging stimulants, make progress on developing a universal flu vaccine, prioritize vector-borne disease research, and support industries of the future".

- *"Advances Vector-Borne Disease Prevention and Control"*. The threat of mosquito and tick-borne diseases continues to rise in the U.S. Cases of tick-borne diseases, such as Lyme disease and Rocky Mountain spotted fever, affected nearly 60,000 Americans in 2017".

- The White House also released a fact sheet titled *"Protecting our Nation's Health and Wellness"*, reinforcing the prioritization of critical health research and advances in vector-borne disease prevention and control. Excerpt: "Section: Advances Vector-Borne Disease Prevention and Control. The Budget includes $66 million for CDC&P's vector- borne disease activities, a $14 million increase compared to the 2020 enacted level that focuses on tick-borne diseases. The Budget also invests in NIH research to improve the Nation's understanding of vector-borne diseases".

Department of Defense FY'20 Tick-Borne Disease Research Program

The program is managed by the (U.S.) Department of Defense, Office of Congressionally Directed Medical Research Programs. Funding opportunities available for the Tick-Borne Disease Research Program includes a Career Development Award and an Idea Development Award.

Contact:
Congressionally Directed Medical Research Programs (CDMRP), Public Affairs
Email: usarmy.detrick.medcom-cdmrp.mbx.cdmrp-public-affairs@mail.mil
Telephone: 301-619-9783

House Lyme Disease Caucus 2020

The (U.S.) House Lyme Disease Caucus is a bi-partisan group working together in Congress to take action on Lyme and tick-borne diseases. It has initiated letters and actions to benefit Lyme patients, such as the inclusion of monies for Lyme and tick-borne diseases into the Congressionally Directed Medical Research Programs (CDMRP), and language and Lyme monies into Appropriations over the years, and initiated favorable legislation. Additionally, it has queried government agencies over policies not favorable to patients.

Advancing vector-borne diseases prevention and control

The 2021 U.S. budget request includes $66 million for CDC&P's vector- borne disease activities, a $14 million increase compared to the 2020 enacted level that focuses on tick-borne diseases (TBD). The Budget also invests in NIH research to improve the Nation's understanding of vector-borne diseases.

The (U.S.) National Institutes of Health (NIH) release their strategic plan for TBDs

Through inventories sent to government agencies to determine gaps in their research on tick-borne diseases (TBDs), the (U.S.) Department of Health & Human Services, Working Group on TBDs, uncovered the fact that NIH did not have a national strategy for TBDs. Building on this report and associated activities, the NIH issued its *"2019 Strategic Plan for Tick-borne Disease Research"*. With input from research and medical communities, patient advocacy groups, pharmaceutical industry, and the general public during its creation and implementation, this plan included (a) all stages of Lyme disease, (b) coordination of research funding across all research agencies to "increase knowledge of pathogenesis and improve diagnosis, and (c) development and testing of new therapeutics for tick-borne diseases. It will be updated every five years.

This newly released NIH plan focuses on six scientific priorities important for advancing research and development over the next five years:

- *Improving fundamental knowledge of TBDs:* This includes:

- The biology of tick-borne pathogens;
- How they are transmitted to humans, evade the immune system, and spread within the body;
- The cause of persistent symptoms in some infected people; and
- Furthering the understanding of how tick-derived factors contribute to the establishment and severity of the disease.

- *Advance research to improve detection and diagnosis of TBDs:* This includes:
  - Improving detection and diagnosis of TBDs;
  - Developing rapid diagnostic tests that can detect a pathogen both early and late in infection;
  - Distinguishing between active and past infections; and
  - Supporting the development of diagnostics capable of predicting treatment success and identifying human biomarkers of infection and persistent symptoms.

- *Accelerate research to improve prevention of TBDs.* The new plan prioritizes:
  - Accelerating research designed to prevent TBD infection;
  - Include vaccines, immune-based treatments, and strategies to reduce the transmission of tick-borne pathogens to animal populations that serve as hosts.

- *Focusing on research:* To develop:
  - New treatments for TBDs; and
  - Techniques to reduce disease complications.

- *Prioritizing the development of tools and resources to advance TBD research:* By:
  - improving scientists' access to biological samples and TBD genetic data; and
  - Supporting preclinical development of promising products.

- *Expanding collaborations across NIH Institutes and Centers:* To promote a :
  - Multidisciplinary approach to TBD research;
  - Answer complex biological questions; and
  - Encourage the application of state-of-the-art technologies used successfully in a range of scientific disciplines.

CDC&P focuses on maternal-fetal transmission of Lyme disease

The mother-to-baby transmission aspect of Lyme disease has long been known but not well publicized. CDC&P recognizes it, emphasizing that *"untreated Lyme disease during pregnancy can lead to infection of the placenta. Spread from mother to fetus is possible but rare. Fortunately, with appropriate antibiotic treatment, there is no increased risk of adverse birth outcomes"*. There are no published studies assessing developmental outcomes of children whose mothers acquired Lyme disease during pregnancy". LymeHope in Canada has been strongly advocating for wider recognition of this

Figure 22.1 – Pregnancy and Lyme disease

Source: CDC&P

aspect. It remains for CDC&P focus to ensure that health departments and physicians are aware of the situation and take appropriate steps to ensure pregnant women get the necessary care". Figure 22.1 reproduces a poster by CDC&P highlighting the relation between pregnancy and Lyme disease.

Canada acknowledges maternal-fetal transmission of Lyme disease

This news pertains specifically to Canadians concerned about the issue of maternal-fetal (*in-utero*) infection also described as transplacental transmission. Lyme disease is not only transmitted through a tick-bite, it can be passed from human-to-human, and mother-to-child in pregnancy.

The World Health Organization (WHO) recognizes congenital Lyme *borreliosis*

The World Health Organization (WHO) has updated its International Classification of Diseases (ICD) to include congenital Lyme *borreliosis* (ICD code # 1C1G2). ICD, now in its eleventh revision, is a system of medical coding created by the WHO for documenting diagnoses, diseases, signs and symptoms, and social circumstances. This official recognition of congenital Lyme as an alternate mode of Lyme transmission comes no less than ~ 35 years after it had been known and documented by multiple international physicians, researchers, scientists, and other experts.

In his article "The enlarging spectrum of tick-borne spirochetes: R.R. Parker Memorial Address" published in the *Reviews of Infectious Diseases Journal*, volume 8, number 6, November-December 1986, Dr. Willy Burgdorferi clearly recognized this fact, stating "*... now we had found a spirochete capable of spreading transplacentally to the organs of the fetus, causing congenital heart disease and possible death of the infant...*".

Take-home points

- On December 20, 2019, President Trump signed the Kay Hagan Tick Act into law to provide funding for the development of a vector-borne disease national strategy, surveillance, and other associated issues. His FY'21 budget includes an increase for vector-borne diseases. The program is managed by the (U.S.) Department of Defense, Office of Congressionally Directed Medical Research Programs.

- The (U.S.) House Lyme Disease Caucus is a bi-partisan group working together in Congress to take action on Lyme and tick-borne diseases. It has initiated letters and actions to benefit Lyme patients.

- The (U.S.) National Institutes of Health (NIH) have released their strategic plan for tick-borne diseases. It focuses on six scientific priorities: Improving fundamental knowledge of tick-borne diseases; advancing research to improve detection and diagnosis of tick-borne diseases (TBDs); accelerating research to improve prevention of TBDs; focusing on research; prioritizing the development of tools and resources to advance TBD research; and expanding collaborations across NIH institutes and centers.

- The Center for Disease Control & Prevention (CDC&P) also focuses on maternal-fetal transmission of Lyme disease, an aspect of the disease that has long been known but not well publicized.

- A leader in maternal-fetal (*in-utero*) infection (also described as transplacental transmission), Canada has long acknowledged and advocated maternal-fetal transmission of Lyme disease.

- The World Health Organization (WHO) has updated its International Classification of Diseases (ICD) to include congenital Lyme *borreliosis* for documenting diagnoses, diseases, signs and symptoms, and social circumstances.

23

Controversies and challenges in treating Lyme and other tick-borne diseases

The material in this Chapter follows in a large part publications by the International Lyme and Associated Diseases Society (ILADS). I provide it in an effort to balance the information provided to patients and to make them aware of certain controversies and challenges in treating their disease.

But, first,...what is ILADS?

The International Lyme and Associated Diseases Society (ILADS) is a non-profit pressure group which advocates for greater recognition and acceptance of the controversial and unrecognized diagnosis "chronic Lyme disease (CLD)".

On ILADS-sustained controversy regarding CLD

ILADS disagrees with mainstream consensus medical views on Lyme disease (LD), leading it to adopt and publish alternative diagnostic criteria and treatment guidelines. Further, it sustains the controversy as to the existence of CLD by advocating for long-term antibiotic treatment. However, the existence of persistent *borrelia* infection is presently not supported by high-quality clinical evidence. In addition, the mainstream medical profession has deemed the use of long-term antibiotics as dangerous and contraindicated.

The consensus of major U.S. medical authorities, including the Infectious Diseases Society of America (IDSA), the American Academy of Neurology (AAN), and the National Institutes of Health (NIH) is careful to:

- "***Distinguish the diagnosis and treatment*** of "*patients who have had well-documented LD and who remain symptomatic for many months to years after completion of appropriate antibiotic therapy from patients who have not had well-documented LD*";

- "***Accept the existence of post–LD symptoms*** in a minority of patients who have had Lyme;

- "***Reject long-term antibiotic treatment*** even for these patients, as entailing too much risk and lacking sufficient efficacy to subject patients to the risks; and

- ***"Call for more research*** into understanding the pathologies that afflict patients with post-Lyme syndrome and into better treatments".

Recapitulating what is LD in a nutshell

LD is a bacterial infection caused by members of the *Borrelia burgdorferi sensu lato* complex. Although the disease was initially identified in Europe, it was (mis)named after the town where the first group of U.S. cases was described. While historically most cases clustered into certain geographic regions, the infection is increasingly widespread across the globe. LD is the most common vector-borne disease in the United States, with the CDC&P estimating that more than 300,000 cases are diagnosed each year. LD is also the most common vector-borne illness in Europe.

LD can be very serious. The infection is often multi-systemic – involving joints, heart, and the nervous system. Although early recognition and treatment lead to resolution of illness for many patients, there are many who live with persistent, debilitating symptoms, and persistent infection – which ILADS terms "chronic Lyme disease" (CLD).

LD and other tick-borne illnesses remain poorly understood, and ILADS finds that guidelines that address the illness in surveillance case terms, including those historically promoted by the CDC&P, do not serve patients well. In the following paragraphs, and without necessarily adopting ILADS' viewpoint, I attempt to explain the many challenges faced by patients with tick-borne illnesses, including controversies over terminology and treatment. I further seek to describe ILADS' position in these controversies and challenges, and ILADS' treatment guidelines.

The controversies in LD exist in a setting of incomplete scientific evidence around tick-borne diseases, including a lack of validated direct testing methods which can be applied across all stages of the disease to accurately distinguish infected from uninfected patients. Further, serologic testing, often held as a gold standard, has significant performance limitations.

Nature of the controversy over diagnosis

LD is a clinical diagnosis depending on three factors:

- *History;*
- *Physical examination;* and
- **Supportive appropriate laboratory testing**.

These three elements of diagnosis are placed in the context of three considerations:

- *Activities and experiences of the patient;*

- *Environmental exposures and risk factors;* and

- *Consideration of other diagnoses that may explain or impact the patient's symptoms.*

At the heart of the controversy is the *relative weight given to each element that contributes to the diagnosis*. ILADS' position, which I espouse, is that no single data point, including serologic testing, automatically outweighs the contributions from all of the data in the patient's presentation and evaluation.

Issues with diagnosis and diagnostics

A distinction must be made between "diagnosis" and "diagnostics". A *diagnosis* encompasses the considered explanation for a patient's symptoms based on the elements described above, as well as the process by which the diagnosis itself is reached. On the other hand, d*iagnostics* refers to laboratory and other test modalities used to aid in reaching a diagnosis. When considering LD, challenges exist in both areas.

Diagnosis

Challenges in diagnosis occur in a setting of varied clinical presentations often appearing in the absence of a history of tick bite. Features that contribute to diagnostic uncertainties and controversy include:

- *Latency of the infection:* Some signs and symptoms do not develop for weeks, months or even years after the time of initial infection.

- *Erythema migrans (EM) rash:* Although a hallmark of the disease, it is not always present or remembered and, in its most common form, does not resemble the classically described "bull's eye" rash.

- *Lyme borreliosis:* The sometimes subtle physical findings and diversity of manifestations of Lyme *borreliosis* require awareness on the part of the examiner in order to discern this infection.

- *Lyme arthritis with effusion:* It has been reported that diagnosis in adults often takes longer than in children.

Diagnostic tests: direct

Direct tests (culture, PCR) for *Borrelia burgdorferi* have performance limitations that must be considered when used in evaluating LD:

- *Availability:* These tests are not readily available;

- ***Growth:*** The slow growth characteristics of *Borrelia burgdorferi* in culture and the sparse distribution of the bacteria in blood and tissues result in generally low sensitivities for these tests;

- ***Cultures and PCR:*** The most successful cultures and PCRs have been on specimens from the leading edge of *erythema migrans* rashes, a manifestation typically not requiring culture or PCR to clarify the diagnosis. When both culture and PCR have been performed on the same specimens, they have not necessarily given the same result. PCR in some studies is shown to perform fairly well in evaluating synovium of Lyme arthritis. In U.S. studies, blood and CSF generally have very low yield on direct tests for *Borrelia burgdorferi*. Differences in testing methods, tissue, and specimens being evaluated contribute to the challenges in direct testing. Emerging direct testing modalities, including a more promising method of culture, present new opportunities for addressing the challenges inherent in direct tests. As in all testing, interpretation of positives and negatives depends upon the clinical setting.

Diagnostic tests: serologic

Serologic tests are the most commonly performed tests for investigating LD. These tests do not identify the presence or absence of *Borrelia burgdorferi,* but rather detect the presence of an antibody response that may be attributable to exposure to this pathogen. Of the dozens of serologic tests available in the U.S. market, none are FDA-approved. Rather, all are FDA-*cleared*, meaning that the tests compare favorably with, and perform at least as well as, other tests already in use, but have not been clinically validated in prospective studies. No test has shown clear superiority in performance.

The reliability of diagnostic testing is a function of accuracy and reproducibility. Accuracy is a test's ability to detect disease when it is present, and not detect disease when it is absent. Reproducibility is the ability of a test to give the same result for a specimen on repeat testing. Highly accurate tests generate few false negatives and few false positives. Unfortunately, Lyme serology produces many false negatives and false positives. Additionally, investigators have repeatedly demonstrated that both ELISA and immunoblot tests for Lyme have poor reproducibility.

Test performance

When antibody testing is done too early or too late in infection, false negatives may result. Further, it has been shown that after an antibody response develops, it can wane or persist regardless of disease status. Finally, some patients fail to develop antibody responses to this infection.

Interpreting test results within the context of the patient's clinical history is important in assessing the likelihood that a positive test represents the presence of the disease and that a negative test represents its absence. The occurrence of false negatives and false positives greatly contributes to the controversy over diagnosis.

Performance characteristics are not only affected by the timing of testing but also by the particular disease manifestation being evaluated. Western blot testing, for example, has been found to perform very well in cases of arthritis (96% sensitivity) but less well in neurologic presentations (72% sensitivity). Note that even when the tests perform well, some cases will be missed if one relies solely on test results. In vaccine trials, reliance on serology alone for the diagnosis of LD would have missed one third of cases.

ELISA

The ELISA quantitatively measures IgM and IgG antibodies or their combination in an automated format. ELISA sensitivities have been shown to vary by disease stage and manifestation. In early Lyme, sensitivities of ~40% have been reported, rising to as high as 82% in disseminated or late/convalescent samples. In a clinical setting consistent with LD, the likelihood that a positive ELISA is a true positive is high. However, in the same setting, the likelihood that a negative ELISA is a true negative is less certain.

Western blot interpretation

Western blots, either IgM or IgG, qualitatively measure antibody reactions across a range of more than two dozen *Borrelia burgdorferi* antigens. The determination of which antibody reactions to include in establishing criteria for a positive Western blot has been the subject of much controversy. There are two current interpretations that have been adopted for standardization:

- *Frequency of particular antibodies appearance:* It has been proposed by Dressler *et al.* in clinically well-characterized patients. Applying statistical parameters, they determined which particular antigens to include and the minimum number of antibody responses needed to achieve a test specificity of 99%. Of interest is the absence of 31 and 34 kDa proteins (OspA and OspB), which may occur infrequently but are nonetheless significant for *Borrelia burgdorferi*, especially when occurring together.

- *Interpretation of IgG Western blot:* Engstrom *et al.*'s have proposed criteria for interpretation of IgM Western blot.

Two-Tier Testing

The commonly recommended testing scheme for evaluating a person suspected of having LD calls for a first step using a highly sensitive ELISA, only followed by a Western Blot if the first step result is positive or equivocal. Negative samples by ELISA are not investigated further. Tests used in this strategy are subject to the confounders of testing described above, and because they are linked sequentially in the stated manner, those confounders are magnified, making the possibility of false negatives much greater.

Like the interpretation parameters of the tests involved, this test strategy was adopted for standardization of testing practices, which were deemed unreliable at the time. Primarily designed and intended to enhance testing specificity, the overall sensitivity of a sequence of this sort necessarily falls. Nevertheless, it was adopted by the CDC&P and the Association of State and Territorial Public Health Laboratory Directors (ASTPHLD) "as a temporary measure", which has not since been replaced.

Clinical practice necessarily places emphasis on identifying every infected patient in order to identify, treat, and prevent adverse consequences. Therefore, ILADS concluded that the two-tiered test strategy does not adequately serve clinician or patient.

The ILADS position on diagnosis and testing

LD is a clinical diagnosis based on history, physical findings, and appropriate supportive laboratory tests when they are indicated. These elements must be considered in the context of the individual patient's full story and with consideration of other diagnoses that may explain or confound the patient's diagnosis. No single element of the diagnostic process outweighs the full and complete evaluation. The strengths and limitations of laboratory testing must be understood by the clinician in order to use testing modalities effectively and avoid some of the pitfalls of diagnosis that can result from over-reliance on laboratory testing to rule-in or rule-out an illness.

Controversy over treatment

LD is a complex illness with many variables. The severity of the infection and a person's response to treatment may depend on :

- *Bacteria species or strains:* Which ones the person has;

- *Illness length:* How long has that person been ill;

- *Body systems involved:* Which ones;

- *Presence of other tick-borne diseases:* and

- *Underlying health status.*

Commonly prescribed antibiotic regimens generally follow a one-size-fits-all approach using 30 or fewer days of antibiotic therapy, but several investigators have found that many people remain ill after receiving such care. In light of all of the potential variables, ILADS maintains that antibiotic treatment must be individualized.

Based on the extensive experience of its membership in the U.S. and across the globe, ILADS members have treated tens of thousands of patients with Lyme and associated tick-borne diseases in the U.S. and across the globe. Based on the collective work of its members, ILADS has a wealth of clinical expertise in treating complex and advanced cases of Lyme disease.

To determine the best possible course of action given the available data, ILADS places a high value on:

- *Preventing chronic infection;*

- *Not causing the abrogation of the immune response;* and

- *Ability of the clinician to exercise clinical judgment.*

As of February 2018, ILADS' treatment guidelines are the only LD Guidelines available at the National Guideline Clearinghouse, an initiative of the DHHS' Agency for Healthcare Research & Quality (AHRQ). These guidelines were developed with careful attention to the available peer-reviewed scientific evidence.

Summary of ILADS' recommendations

Generally speaking, treatment decisions hinge on three factors:

- *Signs and symptoms the person has;*

- *Duration of the illness;* and

- *Previous treatment for this specific illness.*

For treatment, ILADS recommends that:

- *Prophylaxis be discussed with all who have had a black-legged tick bite:* to prevent the onset of infection. (Note: technically speaking, antibiotic prophylaxis of a known bite is not treatment because the person is not yet ill.);

- *Antibiotic prophylaxis applied if decided upon:* 20 days of doxycycline (provided there are no contraindications) with a recommendation against single-dose doxycycline.

- *Antibiotic prophylaxis for erythema migrans:* initial 4-6 weeks course therapy. Subsequent management decisions will be based on whether the signs and symptoms remain or relapse.

- *For persistent (chronic) signs and symptoms of LD:* individualized care that tailors antibiotic treatment to the specific situation. The duration of treatment and the choice of antibiotic or antibiotic combinations are clinical decisions to be made with several factors in mind. Long-term antibiotic therapy is not without risks, and should only proceed under close supervision.

Controversy over CLD

ILADS defines chronic Lyme disease (CLD) as "*...an ongoing infection with any of the pathogenic bacteria in the Borrelia burgdorferi sensu lato group, (that is) poorly understood and often mischaracterized*". Although the infection was often described as *chronic* early in the history of LD, that terminology was abandoned in the late 1990s by the IDSA and many IDSA-affiliated researchers and clinicians. The rationale for that switch is unclear, as the bacteria's ability to cause a chronic and sometimes post-antibiotic persistent infection was already well documented in the scientific literature. The CDC&P generally maintains that LD is an *acute* infection and it does not readily acknowledge that the infection can persist following antibiotic treatment. Instead, the CDC&P recognizes other potential causes for ongoing symptoms, endorsing the use of the term "Post-Treatment Lyme Disease Syndrome"(PLDS) when symptoms persist for more than six months following antibiotic treatment.

ILADS agrees that, for many patients, LD is strictly an acute infection. However, for many other individuals, the infection is chronic. Common symptoms include:

- *Fatigue;*
- *Cognitive dysfunction;*
- *Headaches;*
- *Sleep disturbances;*
- *Migratory myalgia and arthralgia;*
- *Numbness and tingling;*
- *Neuropathic pain;*
- *Depression and anxiety*; and
- *Musculoskeletal problems.*

The diagnostic controversies discussed above also apply to patients with CLD, causing many medical providers to miss a Lyme diagnosis. As a result, patients can exhibit significant symptoms of LD for years and even decades that are misattributed to other entities — fibromyalgia, chronic fatigue syndrome (CFS), and depression are common misdiagnoses. Although Lyme symptoms overlap with the symptoms of these illnesses, a well-conducted meta-analysis demonstrated that persistent LD symptoms were a distinct set of symptoms that differed from those of other chronic, difficult-to-manage diseases. Additionally, in comparing the cerebrospinal fluid of patients with chronic, post-treatment manifestations of LD to that of patients with CFS, researchers found that the two groups could be distinguished from each other on the basis of proteins unique to a particular group.

Patients with CLD may require prolonged treatment — i.e., additional courses of antibiotics. An unknown percentage of patients (10% to more than 30%) report prolonged symptoms despite prior antibiotic treatment for LD. Given *Borrelia burgdorferi*'s wide array of survival mechanisms, this statistic is not surprising. As discussed earlier in this book, *Borrelia burgdorferi* is notoriously adaptable, which enhances its survival. The bacteria can alter its morphology, engage in antigenic variation, invoke periods of dormancy, inhabit protective niches, and form biofilms that protect the bacteria from the immune system and the effects of antibiotics. These survival mechanisms allow for

Borrelia burgdorferi persistence following short-term antibiotic therapy. Repeated courses of antibiotics that do not address these bacterial protections are also likely to fail.

ILADS position on treating CLD

Based on the extensive collective experience of ILADS members and a rigorous review of the broader scientific literature, ILADS maintains it is in the best interest of CLD patients for clinicians to offer additional treatment. Taking into account the strength of the evidence addressing the effectiveness of antibiotic re-treatment, the burden of disease in this patient population, and the risks associated with various antibiotic options, ILADS concludes that the very real consequences of an untreated chronic Lyme infection far outweigh the potential consequences of long-term antibiotic therapy. Further, it asserts that although it is too early to standardize restrictive protocols, effective treatment options are available for these patients.

Controversy over time of attachment before transmission

Many professionals mistakenly believe that LD cannot be transmitted when a black-legged tick has been attached for 24 to 48 hours. This is incorrect. Several animal studies have documented that although the risk of LD is quite low for attachment times under 24 hours, the risk is not zero. The length of attachment for transmission of other members of the *Borrelia burgdorferi sensu lato* complex pathogens to occur may differ from that of *Borrelia burgdorferi*. ILADS recommends that patients with a known black-legged tick bite of any duration should discuss the possibility of LD with their providers.

Other challenges in treating LD

Other tick-borne infections

Black-legged ticks can transmit at least four other infections. The frequency of tick-borne co-infections in LD patients from endemic areas ranges from 4%-45%. Not all providers appreciate the risks of acquiring co-infections nor the clinical significance of these pathogens. If co-infections are unidentified and unaddressed, a provider may treat LD effectively, yet have patients who remain ill.

Pressures on clinicians

Medical providers face the challenge of these often complicated patients in practice settings that may impose time constraints and other limitations to their choices in diagnosis and treatment. The diagnostic possibilities are themselves complex and the requirements for follow-up, careful observation of progress, and further decision-making significant. Placed in the context of a "controversial illness" and personal concerns about experience and expertise in this area of medicine, the pressure on individual clinicians can be substantial.

Misuse of treatment guidelines

To inform patient care, treatment guidelines were called for by the Institute of Medicine (IOM). These were meant to assist clinicians by assembling the best available evidence from the medical literature as well as assessments of risks and benefits of treatments and alternatives. The IOM standards for guidelines development recommend that patients and others affected by a guideline be included in the development process as this would help assure that guidelines address and prioritize the issues most important to patients. Guidelines are not intended to supplant the judgment of a clinician in the care of individual patients. When insurers, employers, medical boards, or other entities treat guidelines as if they are laws, misusing them to determine payment, as performance measures, or for disciplinary purposes, it can undermine the utility of these works for clinicians.

Insurance issues

Health insurance companies create policies for payment of benefits to guide in claims decisions. When these policies are narrowly determined, as they are when based on surveillance case definitions of LD, they may impede a patient's recovery of benefits, and importantly, may prevent receipt of care. When cost assessments are performed, ILADS holds that they need to include the costs of:

- *Non-treatment;*

- *Other kinds of management:* that will be required if the underlying cause of illness is not treated; and

- *Disability:* as a result of the illness.

Marginalization of patients with LD

While any patient, particularly one with a chronic illness, is at risk of marginalization, LD patients face a particularly intense set of challenges:

- *Stigma associated with this controversial diagnosis;*

- *Polarization of medical providers concerning diagnosis and treatment;*

- *Challenges of obtaining insurance coverage:* for treatment beyond the acute phase of the disease; and

- *Potential to experience isolation because of the illness.*

All of these factors and the controversies explored above contribute to the experiences of patients who struggle with the very real effects of LD. The medical uncertainties surrounding tick-borne diseases (TBD) may lead some clinicians to take a hands-off approach to these illnesses, or to fail to coordinate with a Lyme-treating colleague. Patients experience misdiagnoses and, at times, their symptoms may be erroneously attributed to a psychiatric diagnosis or malingering. The cumulative effect is that many patients become marginalized.

The evidence regarding Lyme and other tick-borne diseases continues to evolve. Many in the greater medical community may not be sufficiently versed in the scientific evidence of LD and other TBDs, giving rise to misconceptions about the disease. Evidence-based medicine is the integration of clinical research, clinical expertise, and patient values and preferences. Given that the clinical evidence is limited and of low quality, clinical expertise and patient values take on increased importance. To assure the decisions about treatment and outcomes remain patient-centered and individualized, ILADS strongly encourages patients to be actively involved in decisions affecting their medical care.

Take-home points

- The consensus of major U.S. medical authorities consists in: distinguishing the diagnosis and treatment of patients who have had well-documented Lyme disease (LD) and who remain symptomatic for many months to years after completion of appropriate antibiotic therapy from patients who have not had well-documented LD; accepting the existence of post–LD symptoms in a minority of patients who have had Lyme; rejecting long-term antibiotic treatment even for these patients, as entailing too much risk and lacking sufficient efficacy to subject patients to the risks; and calling for more research into understanding the pathologies that afflict patients with post-Lyme syndrome.

- The International Lyme and Associated Diseases Society (ILADS) disagrees with mainstream consensus medical views on LD, leading it to adopt and publish alternative diagnostic criteria and treatment guidelines. Further, it sustains the controversy as to the existence of chronic Lyme disease (CLD) by advocating for long-term antibiotic treatment.

- There exists a controversy over the diagnosis. LD as a clinical diagnosis depending on three factors (history; physical examination; and supportive appropriate laboratory testing) placed in the context of three considerations (activities and experiences of the patient; environmental exposures and risk factors; and consideration of other diagnoses). At the heart of the controversy is the relative weight given to each element that contributes to the diagnosis. No single data point, including serologic testing, should automatically outweigh the others.

- Challenges in diagnosis occur in a setting of varied clinical presentations often appearing in the absence of a history of tick bite. Features that contribute to diagnostic uncertainties and controversy concern the latency of infection, *erythema migrans* (EM) rash, Lyme *borreliosis*, and Lyme arthritis with effusion.

- Other challenges occur with the diagnostic tests (ELISA, Western immunoblots) regarding availability, growth, cultures, serology, reliability, and performance. The commonly recommended two-tier testing scheme calls for a first step using a highly sensitive ELISA, only followed by a Western Blot if the first step result is positive or equivocal. Negative samples by ELISA are not investigated further. It was adopted by the CDC&P and other governmental entities "as a temporary measure", which has not since been replaced although it may not adequately serve the clinician or the patient.

- There is further controversy over treatment. Commonly prescribed antibiotic regimens generally follow a one-size-fits-all approach using 30 or fewer days of antibiotic therapy but many people remain ill after receiving such care, calling for individualized antibiotic treatment.

- To determine the best possible course of action given the available data, a high value is placed on preventing chronic infection; not causing the abrogation of the immune response; and not impairing the ability of the clinician to exercise clinical judgment.

- Treatment decisions hinge on three factors (signs and symptoms; duration of the illness; and previous treatment). For treatment, ILADS recommends antibiotic prophylaxis (20 days of doxycycline barring contraindications); not a single-dose of doxycycline. For *erythema migrans,* antibiotic prophylaxis (initially 4-6 weeks course therapy) and individualized care.

- There is an additional controversy regarding chronic Lyme disease (CLD), which is an ongoing infection with any of the pathogenic bacteria in the *Borrelia burgdorferi sensu lato* group that is poorly understood and often mischaracterized.

- Common CLD symptoms include fatigue; cognitive dysfunction; headaches; sleep disturbances; migratory myalgia and arthralgia; numbness and tingling; neuropathic pain; depression and anxiety; and musculoskeletal problems.

- The above diagnostic controversies also apply to patients with CLD, causing many medical providers to miss a Lyme diagnosis. As a result, patients can exhibit significant symptoms of LD for years and even decades that are misattributed to other entities (fibromyalgia, chronic fatigue syndrome, and depression). Patients with CLD may require prolonged treatment — i.e., additional courses of antibiotics.

- *Borrelia burgdorferi* is notoriously adaptable, which enhances its survival. It can alter its morphology, engage in antigenic variation, invoke periods of dormancy, inhabit protective niches, and form biofilms that protect it from the immune system and the effects of antibiotics. These survival mechanisms allow it to persist following short-term antibiotic therapy. Repeated courses of antibiotics that do not address these bacterial protections are also likely to fail.

- There exists a further controversy over the tick's time of attachment before disease transmission. The view held by many professionals that LD cannot be transmitted when a black-legged tick has been attached for 24-48 hours is incorrect.

- There are other challenges in treating LD, including other tick-borne infections; pressures on clinicians (complex diagnostic possibilities and requirements for follow-up; careful observation of progress; and further decision-making); misuse of treatment guidelines; insurance issues (including non-treatment; management; and disability); and marginalization of patients (stigma associated with this controversial diagnosis; polarization of medical providers; challenges of obtaining insurance coverage; and potential to experience isolation because of the illness). All of these factors contribute to the experiences of patients who struggle with the very real effects of LD.

- The evidence regarding Lyme and other tick-borne diseases continues to evolve. Many in the greater medical community may not be sufficiently versed in the scientific evidence of LD and other tick-borne diseases, giving rise to misconceptions about the disease. Patients are strongly encouraged to be actively involved in decisions affecting their medical care.

References

Agger WA, Callister SM, and Jobe DA (1992). "*In vitro* susceptibilities of *Borrelia burgdorferi* to five oral cephalosporins and ceftriaxone". *Antimicrob Agents Chemother* **36**(8):1788-90.

Agre F and Schwartz R (1993). "The value of early treatment of deer tick bites for the prevention of Lyme disease". *Am J Dis Child* **147**(9):945-7.

Aguero-Rosenfeld ME, Nowakowski J, and Bittker S, (1996). "Evolution of the serologic response to *Borrelia burgdorferi* in treated patients with culture-confirmed erythema migrans". *J Clin Microbiol* **34**(1):1-9.

Albert S, Schulze J, Riegel H, and Brade V (1999). "Lyme arthritis in a 12-year-old patient after a latency period of 5 years". *Infection* **27**(4-5):286-8.

Alder J, Mitten M, Jarvis K, *et al.* (1993). "Efficacy of clarithromycin for treatment of experimental Lyme disease *in vivo*". *Antimicrob Agents Chemother* **37**(6):1329-33.

Al-Robaiy S, Dihazi H, Kacza J, *et al.* (2010). "Metamorphosis of *Borrelia burgdorferi* organisms - RNA, lipid and protein composition in context with the spirochetes' shape". *J Basic Microbiol* **50**(Suppl 1):S5-17.

Altman DG (2009). "Missing outcomes in randomized trials: addressing the dilemma". *Open Med* **3**(2):51-3.

Asch ES, Bujak DI, Weiss M, *et al.* (1994). "Lyme disease: an infectious and post-infectious syndrome". *J Rheumatol* **21**(3):454-61.

Aucott J, Morrison C, Munoz B, *et al.* (2009). "Diagnostic challenges of early Lyme disease: lessons from a community case series". *BMC Infect Dis* **9**:79.

Aucott JN, Rebman AW, Crowder LA, Kortte KB (2013). "Post-treatment Lyme disease syndrome symptomatology and the impact on life functioning: is there something here?" *Qual Life Res* **22**(1):75-84.

Auwaerter PG (2007). "Point: antibiotic therapy is not the answer for patients with persisting symptoms attributable to lyme disease". *Clin Infect Dis* **45**(2):143-8.

Bacon RM, Kugeler KJ, and Mead PS. (2008). "Surveillance for Lyme disease--United States, 1992-2006". *MMWR Surveill Summ* **57**(10):1-9.

Barbour AG and Restrepo BI (2000). "Antigenic variation in vector-borne pathogens". *Emerg Infect Dis* **6**(5):449-57.

Barsic B, Maretic T, Majerus L, and Strugar J. (2000). "Comparison of azithromycin and doxycycline in the treatment of erythema migrans". *Infection* **28**(3):153-6.

Barthold SW, Hodzic E, Imai DM, *et al.* (2010). "Ineffectiveness of tigecycline against persistent *Borrelia burgdorferi*". *Antimicrob Agents Chemother* **54**(2):643-51.

Berger BW (1988). "Treatment of *erythema chronicum migrans* of Lyme disease". *Ann N Y Acad*

Sci **539**:346-51.

Berger BW (1989). "Dermatologic manifestations of Lyme disease". *Rev Infect Dis* **11**(Suppl 6):S1475-81.

Borg R, Dotevall L, Hagberg L, et al. (2005). "Intravenous ceftriaxone compared with oral doxycycline for the treatment of Lyme neuroborreliosis". *Scand J Infect Dis* 37(6-7):449-54 [Taylor & Francis Online].

Bradley JF, Johnson RC, and Goodman JL (1994). "The persistence of spirochetal nucleic acids in active Lyme arthritis". *Ann Intern Med* **120**(6):487-9.

Brouqui P, Badiaga S, and Raoult D (1996). "Eucaryotic cells protect *Borrelia burgdorferi* from the action of penicillin and ceftriaxone but not from the action of doxycycline and erythromycin". *Antimicrob Agents Chemother* **40**(6):1552-4.

Cabello FC, Godfrey HP, and Newman SA (2007). "Hidden in plain sight: *Borrelia burgdorferi* and the extracellular matrix". *Trends Microbiol* **15**(8):350-4.

Cairns V and Godwin J. (2005). "Post-Lyme *borreliosis* syndrome: a meta-analysis of reported s ymptoms". *Int J Epidemiol* **34**(6):1340-5.

Cameron DJ (2007). "Consequences of treatment delay in Lyme disease". *J Eval Clin Pract* **13**(3):470-2.

Cameron DJ (2008). " Severity of Lyme disease with persistent symptoms: insights from a double-blind placebo-controlled clinical trial". *Minerva Med* **99**(5):489-96.

Cerar D, Cerar T, Ruzic-Sabljic E, et al. (2010). "Subjective symptoms after treatment of early Lyme disease". *Am J Med* **123**(1):79-86.

Cimmino MA and Accardo S (1992). "Long term treatment of chronic Lyme arthritis with benzathine penicillin". *Ann Rheum Dis* **51**(8):1007-8.

Cimmino MA, Moggiana GL, Parisi M, and Accardo S (1996). "Treatment of Lyme arthritis. I nfection" **24**(1):91-3.

Clark RP and Hu LT. (2008). "Prevention of Lyme disease and other tick-borne infections". *Infect Dis Clin North Am* **22**(3):381-96.

Cooper C (2014). " Safety of long-term therapy with penicillin and penicillin derivatives. Center for Drug Evaluation and Research". Available from: www.fda.gov/Drugs/EmergencyPreparedness/BioterrorismandDrugPreparedness/ucm072755.ht

Corapi KM, White MI, Phillips CB, et al. (2007). "Strategies for primary and secondary prevention of Lyme disease". *Nat Clin Pract Rheumatol* **3**(1):20-5.

Costello CM, Steere AC, Pinkerton RE, and Feder HM Jr (1989). "A prospective study of tick bites in an endemic area for Lyme disease". *J Infect Dis* **159**(1):136-9.

Coyle PK and Schutzer SE. (1991). "Neurologic presentations in Lyme disease". *Hosp Pract* **26**(11):55-66.discussion 66, 69-70 [Taylor & Francis Online].

Cunha BA (2000). "Minocycline *versus* doxycycline in the treatment of Lyme *neuroborreliosis*". *Clin Infect Dis* **30**(1):237-8.

Dattwyler RJ, Halperin JJ, Pass H, and Luft BJ (1987). "Ceftriaxone as effective therapy in refractory Lyme disease". *J Infect Dis* *155*(6):1322-5.

Dattwyler RJ, Halperin JJ, Volkman DJ, and Luft BJ. (1988a). "Treatment of late Lyme borreliosis--randomised comparison of ceftriaxone and penicillin". *The Lancet* **1**(8596):1191-4.

Dattwyler RJ, Volkman DJ, Luft BJ, et al. (1988b). "Seronegative Lyme disease. Dissociation of specific T- and B-lymphocyte responses to *Borrelia burgdorferi*". *N Engl J Med* **319**(22):1441-6.

Dattwyler RJ, Volkman DJ, Conaty S M, et al. (1990). "Amoxycillin plus probenecid versus doxycycline for treatment of *erythema migrans borreliosis*". *The Lancet* **336**(8728):1404-6.

Dattwyler RJ, Grunwaldt E, and Luft BJ (1996). "Clarithromycin in treatment of early Lyme disease: a pilot study". *Antimicrob Agents Chemother* **40**(2):468-9.

Dattwyler RJ, Luft BJ, Kunkel MJ, et al. (1997). "Ceftriaxone compared with doxycycline for the treatment of acute disseminated Lyme disease". *N Engl J Med* **337**(5):289-94.

Delong AK, Blossom B, Maloney EL, and Phillips SE (2012). "Antibiotic retreatment of Lyme disease in patients with persistent symptoms: a biostatistical review of randomized, placebo-controlled, clinical trials". *Contemp Clin Trials* **33**(6):1132-42.

Donta ST (1996). "Macrolide therapy of chronic Lyme disease". *Med Sci Monit* **9**(11):PI136-42.

Donta ST (1997). "Tetracycline therapy for chronic Lyme disease". Clin Infect Dis **25**(Suppl 1):S52-6.

Donta ST (2012). "Issues in the diagnosis and treatment of Lyme disease. *Open Neurol J* **6**:140-5.

Dotevall L and Hagberg L. (1999). "Successful oral doxycycline treatment of Lyme disease-associated facial palsy and meningitis". *Clin Infect Dis* **28**(3):569-74.

Duray PH (1989). "Clinical pathologic correlations of Lyme disease". *Rev Infect Dis* **11**(Suppl 6):S1487-93.

Duray PH, Yin SR, Ito Y, et al.(2005). "Invasion of human tissue *ex vivo* by *Borrelia burgdorferi*". *J Infect Dis* **191**(10):1747-54.

Eikeland R, Mygland A, Herlofson K, and Ljostad U. (2011). "European *neuroborreliosis*: quality of life 30 months after treatment". *Acta Neurol Scand* **124**(5):349-54.

Embers ME, Ramamoorthy R, and Philipp MT (2004). "Survival strategies of *Borrelia burgdorferi*, the etiologic agent of Lyme disease". *Microbes Infect* **6**(3):312-18.

Eppes SC and Childs JA (2012). "Comparative study of cefuroxime axetil versus amoxicillin in children with early Lyme disease". *Pediatrics* **109**(6):1173-7.

Fallon BA, Keilp JG, Corbera KM, et al. (2008). "A randomized, placebo-controlled trial of repeated IV antibiotic therapy for Lyme encephalopathy". *Neurology* **70**(13):992-1003.

Fallon BA, Levin ES, Schweitzer PJ, Hardesty D (2010). "Inflammation and central nervous system Lyme disease". *Neurobiol Dis* **37**(3):534-41.

Fallon BA, Petkova E, Keilp JG, Britton CB (2012). "A reappraisal of the U.S. clinical trials of post-treatment Lyme disease syndrome". *Open Neurol J* **6**:79-87.

Frank C, Fix AD, Pena CA, and Strickland GT (2002). "Mapping Lyme disease incidence for diagnostic and preventive decisions", Maryland. *Emerg Infect Dis* **8**(4):427-9.

Fymat AL (2017a). "Parkinson's disease and other movement disorders: a review", *Journal of Current Opinions in Neurological Science* **2**(1):316-43.

Fymat AL (2017b). "Neurological disorders and the blood-brain barrier: 2. Parkinson's disease and other movement disorders", *Journal of Current Opinions in Neurological Science* **2**(1)362-83.

Fymat AL (2018a). "Blood-brain barrier permeability and neurological diseases", *Journal of Current Opinions in Neurological Science* (Editorial).**2**(2):411-4.

Fymat AL (2018b). "Alzheimer's disease: a review", *Journal of Current Opinions in Neurological Science* **2**(2);415-36,

Fymat AL (2018c). "Regulating the brain's autoimmune system: the end of all neurological disorders?" *Journal of Current Opinions in Neurological Science* **2**(3):475-9.

Fymat AL (2018d). "Alzheimer's disease: prevention, delay, minimization and reversal", *Journal of Clinical Research in Neurology* **1**(1):1-16.

Fymat AL (2018e). "Harnessing the immune system to treat cancers and neurodegenerative diseases",

Journal of Clinical Research in Neurology **1**(1):1-14.

Fymat AL (2018f). "Is Alzheimer's an autoimmune disease gone rogue"", *Journal of Clinical Research in Neurology* **2**(1):1-4.

Fymat AL (2018g). "Dementia treatment: where do we stand?", *Journal of Current Opinions in Neurological Science* **3**(1):1-3.

Fymat AL (2018h). "On dementia and other cognitive disorders", *Journal of Clinical Research in Neurology* **1**(2):1-14.

Fymat AL (2018i). "Is Alzheimer's a runaway autoimmune Disease? and how to cure it?", *Newsletter European Union Academy of Sciences Annual Report (2018).*

Fymat AL (2019a). "Is Alzheimer's a runaway autoimmune disease? and how to cure it?" *Proceedings of the European Union Academy of Sciences*, Newsletter, pages 379-83.

Fymat AL (2019b). "Dementia: a review", *Journal of Clinical Psychiatry and Neuroscience* **1**(3):27-34.

Fymat AL (2019c). "The pathogenic brain", *Journal of Current Opinions in Neurological Science* **3**(2);669-71.

Fymat AL (2019d). "On the pathogenic hypothesis of neurodegenerative diseases", *Journal of Clinical Research in Neurology* **2**(1):1-7.

Fymat AL (2019e). "Dementia with Lewy bodies: a review", *Journal of Current Opinions in Neurological Science* **4**(1);15-32.

Fymat AL (2019f). "Our two interacting brains: etiologic modulations of neurodegenerative and gastroenteric diseases", *Journal of Current Opinions in Neurological Science* **4**(2):50-4.

Fymat AL (2019g). "What do we know about Lewy body dementias?" *Journal of Psychiatry and Psychotherapy (Editorial)* **2**(1)-013:1-4. doi:10.31579/JPP.2019/018.

Fymat AL (2019h). "Viruses in the brain...? any connections to Parkinson's and other neurodegenerative diseases?" *Proceedings of the European Union Academy of Sciences, 2019 Newsletter.*

Fymat AL (2019i). "Alzhei … Who? demystifying the disease and what you can do about it", Tellwell Talent Publishers, pp 235, 2019. ISBN: 978-0-2288-2420-6 (Hardcover); 978-0-2288-2419-0 (Paperback).

Fymat AL (2020a). "Recent research developments in Parkinson's disease", *Current Opinions in Neurological Science* **5**(1):12-30.

Fymat AL (2020b). "The role of radiological imaging in neurodegenerative disorders", *Journal of Radiology and Imaging Science* **1**(1):1-14.

Fymat AL (2020c). "Parkin...ss...oo...nn: elucidating the disease and what you can do about it", Tellwell Talent Publishers, 2020.

Gerber MA, Zemel LS, and Shapiro ED. (1998). "Lyme arthritis in children: clinical epidemiology and long-term outcomes". *Pediatrics* **102**(4 Pt 1):905-8.

Hartiala P, Hytonen J, Suhonen J, *et al.* 2008). "*Borrelia burgdorferi* inhibits human neutrophil functions". *Microbes Infect* **10**(1):60-8.

Haupl T, Hahn G, Rittig M, *et al.* (1993). "Persistence of *Borrelia burgdorferi* in ligamentous tissue from a patient with chronic *Lyme borreliosis*". *Arthritis Rheum* **36**(11):1621-6.

Hayes E. (2003). "Lyme disease". *Clin Evid* (10):887-99.

Hodzic E, Feng S, Freet KJ, and Barthold SW (2003). "*Borrelia burgdorferi* population dynamics and prototype gene expression during infection of immunocompetent and immunodeficient mice". *Infect Immun* **71**(9):5042-55.

Hodzic E, Feng S, Holden K, et al. (2008). "Persistence of *Borrelia burgdorferi* following antibiotic treatment in mice". *Antimicrob Agents Chemother* **52**(5):1728-36.

Hunfeld KP, Weigand J, Wichelhaus TA, et al. (2001). "In vitro activity of mezlocillin, meropenem, aztreonam, vancomycin, teicoplanin, ribostamycin and fusidic acid against *Borrelia burgdorferi*". *Int J Antimicrob Agents* **17**(3):203-8.

Institute of Medicine, (US) Committee on Lyme Disease and Other Tick-Borne Diseases (2011): "The state of the science". In: "Critical needs and gaps in understanding prevention, amelioration, and resolution of Lyme and other tick-borne diseases: the short-term and long-term outcomes: workshop report". *National Academies Press;* Washington, DC, USA.

Jones KD, Burckhardt CS, Deodhar AA, et al. (2008). "A six-month randomized controlled trial of exercise and pyridostigmine in the treatment of fibromyalgia". *Arthritis Rheum* **58**(2):612-22.

Johnson L, Aylward A, and Stricker RB (2011). "Healthcare access and burden of care for patients with Lyme disease: a large United States survey". Health Policy **102**(1):64-71.

Johnson L, Wilcox S, Mankoff J, and Stricker RB (2014). "Severity of chronic Lyme disease compared to other chronic conditions: a quality of life survey". *Peer J* **2**:e322.

Johnson RC, Kodner CB, Jurkovich PJ, and Collins JJ (1990). "Comparative *in vitro* and *in vivo* susceptibilities of the Lyme disease spirochete *Borrelia burgdorferi* to cefuroxime and other antimicrobial agents". *Antimicrob Agents Chemother* **34**(11):2133-6.

Kersten A, Poitschek C, Rauch S, and Aberer E (1995). "Effects of penicillin, ceftriaxone, and doxycycline on morphology of *Borrelia burgdorferi*". *Antimicrob Agents Chemother* **39**(5):1127-33.

Klempner MS, Noring R, and Rogers RA (1993). "Invasion of human skin fibroblasts by the Lyme disease spirochete, *Borrelia burgdorferi*". *J Infect Dis* **167**(5):1074-81.

Klempner MS, Hu LT, Evans J, et al. (2001). "Two controlled trials of antibiotic treatment in patients with persistent symptoms and a history of Lyme disease". *N Engl J Med* **345**(2):85-92.

Kraiczy P, Skerka C, Kirschfink M, et al. (2002). "Immune evasion of *Borrelia burgdorferi*: insufficient killing of the pathogens by complement and antibody". *Int J Med Microbiol* **291**(Suppl 33):141-6.

Krause PJ, Telford SR 3rd, Spielman A, et al. (1996). "Concurrent Lyme disease and *babesiosis:* evidence for increased severity and duration of illness". *JAMA* **275**(21):1657-60.

Krupp LB, Hyman LG, Grimson R, et al. (2003). "Study and treatment of post-Lyme disease (STOP-LD): a randomized double masked clinical trial". *Neurology* **60**(12):1923-30.

Lawrence C, Lipton RB, Lowy FD, and Coyle PK (2012). "Seronegative chronic relapsing *neuroborreliosis*". Eur Neurol **35**(2):113-17.

Lee J and Wormser GP (2008). " Pharmacodynamics of doxycycline for chemoprophylaxis of Lyme disease: preliminary findings and possible implications for other antimicrobials". *Int J Antimicrob Agents* **31**(3):235-9.

Liang FT, Jacobs MB, Bowers LC, and Philipp MT (2002). "An immune evasion mechanism for spirochetal persistence in Lyme *borreliosis*". *J Exp Med* **195**(4):415-22.

Liegner KB, Shapiro JR, Ramsay D, et al. (1993). "Recurrent *erythema migrans* despite extended antibiotic treatment with minocycline in a patient with persisting *Borrelia burgdorferi* infection". *J Am Acad Dermatol* **28**(2 Pt 2):312-4.

Livengood JA and Gilmore RD Jr. (2006). "Invasion of human neuronal and glial cells by an infectious strain of *Borrelia burgdorferi*". *Microbes Infect* **8**(14-15):2832-40.

Lo R, Menzies DJ, Archer H, and Cohen TJ. (2003). "Complete heart block due to Lyme carditis". *J*

Invasive Cardiol **15**(6):367-9.

Logar M, Ruzic-Sabljic E, Maraspin V, et al. (2004). "Comparison of *erythema migrans* caused by *Borrelia afzelii* and *Borrelia garinii*." *Infection* **32**(1):15-9.

Logigian EL, Kaplan RF, and Steere AC. (1990). "Chronic neurologic manifestations of Lyme disease". *N Engl J Med* **323**(21):1438-44.

Logigian EL, Kaplan RF, and Steere AC. (1999). "Successful treatment of Lyme encephalopathy with intravenous ceftriaxone". *J Infect Dis* **180**(2):377-83.

Luft BJ, Dattwyler RJ, Johnson RC, et al. (1996). "Azithromycin compared with amoxicillin in the treatment of *erythema migrans:* a double-blind, randomized, controlled trial". *Ann Intern Med* **124**(9):785-91.

Luger SW, Paparone P, Wormser GP, et al. (1995). "Comparison of cefuroxime axetil and doxycycline in treatment of patients with early Lyme disease associated with *erythema migrans*". *Antimicrob Agents Chemother* **39**(3):661-7.

Maes E, Lecomte P, and Ray N (1998). "A cost-of-illness study of Lyme disease in the United States". *Clin Ther* **20**(5):993-1008.

Maki DG, Kluger DM, and Crnich CJ (2006). "The risk of bloodstream infection in adults with different intravascular devices: a systematic review of 200 published prospective studies". *Mayo Clin Proc* **81**(9):1159-71.

Maloney EL. (2011). "The management of *Ixodes scapularis* bites in the upper Midwest". *WMJ* **110**(2):78-81.

Maraspin V, Lotric-Furlan S, Cimperman J, et al.(1999). *"Erythema migrans* in the immunocompromised host". *Wien Klin Wochenschr* **111**(22-23):923-32.

Marques A, Telford SR 3rd, Turk SP, et al. (2014). "Xenodiagnosis to detect *Borrelia burgdorferi* infection: a first in human study". *Clin Infect Dis* **58**(7):937-45.

Massarotti EM, Luger SW, Rahn DW, et al. (1992). "Treatment of early Lyme disease". *Am J Med* **92**(4):396-403.

Meltzer MI, Dennis DT, and Orloski KA. (1999). "The cost effectiveness of vaccinating against Lyme disease". *Emerg Infect Dis* **5**(3):321-8.

Mullegger RR, Millner MM, Stanek G, and Spork KD (1991). "Penicillin G sodium and ceftriaxone in the treatment of *neuroborreliosis* in children: a prospective study". *Infection* **19**(4):279-83.

Mursic VP, Wanner G, Reinhardt S, et al. (1996). "Formation and cultivation of *Borrelia burgdorferi* spheroplast-L-form variants". *Infection* **24**(3):218-26.

Nadelman RB, Luger SW, Frank E, et al. (1992). "Comparison of cefuroxime axetil and doxycycline in the treatment of early Lyme disease". *Ann Intern Med* **117**(4):273-80.

Nadelman RB, Nowakowski J, Fish D, et al. (2001). "Prophylaxis with single-dose doxycycline for the prevention of Lyme disease after an *Ixodes scapularis* tick bite". *N Engl J Med* **345**(2):79-84.

Nadelman RB, Hanincova K, Mukherjee P, et al. (2012). "Differentiation of reinfection from relapse in recurrent Lyme disease". *N Engl J Med* **367**(20):1883-90.

Nocton JJ, Dressler F, Rutledge BJ, et al. (1994). "Detection of *Borrelia burgdorferi* DNA by polymerase chain reaction in synovial fluid from patients with Lyme arthritis". *N Engl J Med* **330**(4):229-34.

Nowakowski J, McKenna D, Nadelman RB, et al. (2000). "Failure of treatment with cephalexin for Lyme disease". *Arch Fam Med* **9**(6):563-7.

Oksi J, Nikoskelainen J, and Viljanen MK. (1998). "Comparison of oral cefixime and intravenous ceftriaxone followed by oral amoxicillin in disseminated Lyme *borreliosis*". *Eur J Clin*

Microbiol Infect Dis **17**(10):715-9.

Oksi J, Marjamaki M, Nikoskelainen J, and Viljanen MK (1999). "*Borrelia burgdorferi* detected by culture and PCR in clinical relapse of disseminated Lyme *borreliosis*". *Ann Med* **31**(3):225-32 [Taylor & Francis Online].

Oksi J, Nikoskelainen J, Hiekkanen H, et al. (2007). "Duration of antibiotic treatment in disseminated Lyme *borreliosis*: a double-blind, randomized, placebo-controlled, multicenter clinical study". *Eur J Clin Microbiol Infect Dis* **26**(8):571-81.

Pfister HW, Preac-Mursic V, Wilske B, et al.(1991). "Randomized comparison of ceftriaxone and cefotaxime in Lyme *neuroborreliosis*". *J Infect Dis* **163**(2):311-18.

Piesman J, Dolan MC, Happ C M, et al. (1997). "Duration of immunity to reinfection with tick-transmitted *Borrelia burgdorferi* in naturally infected mice". *Infect Immun* **65**(10):4043-7.

Piesman J, Hojgaard A (2012). "Protective value of prophylactic antibiotic treatment of tick bite for Lyme disease prevention: an animal model ticks". *Tick Borne Dis* **3**(3):193-6.

Pollina DA, Sliwinski M, Squires NK, and Krupp LB (1999). "Cognitive processing speed in Lyme disease". *Neuropsychiatry Neuropsychol Behav Neurol* **12**(1):72-8.

Preac-Mursic V, Weber K, Pfister HW, et al. (1989). "Survival of *Borrelia burgdorferi* in antibiotically treated patients with Lyme *borreliosis*". *Infection* **17**(6):355-9.

Preac-Mursic V, Pfister HW, Spiegel H, et al. (1993). "First isolation of *Borrelia burgdorferi* from an iris biopsy". *J Clin Neuroophthalmol* **13**(3):155-61.

Rahn DW (1991). "Lyme disease: clinical manifestations, diagnosis, and treatment. *Semin Arthritis Rheum* **20**(4):201-18.

Reported cases of Lyme disease by year, United States, 1995-2009. Available from: www.cdc.gov/lyme/stats/chartstables/casesbyyear.html.

Sapi E, Bastian SL, Mpoy C M, et al. (2012). "Characterization of biofilm formation by *Borrelia burgdorferi in vitro*". *PLoS One* **7**(10):e48277.

Sartakova ML, Dobrikova EY, Terekhova D A, et al. (2003). "Novel antibiotic-resistance markers in pGK12-derived vectors for *Borrelia burgdorferi*". *Gene* **303**:131-7.

Schaefer C, Chandran A, Hufstader M, et al. (2011). "The comparative burden of mild, moderate and severe fibromyalgia: results from a cross-sectional survey in the United States". *Health Qual Life Outcomes* **9**(1):71.

Schmidli J, Hunziker T, Moesli P, Schaad UB (1988). "Cultivation of *Borrelia burgdorferi* from joint fluid three months after treatment of facial palsy due to Lyme *borreliosis*". *J Infect Dis* **158**(4):905-6.

Schoen RT, Aversa JM, Rahn DW, and Steere AC (1991). "Treatment of refractory chronic Lyme arthritis with arthroscopic synovectomy". *Arthritis Rheum* **34**(8):1056-60.

Schwan TG and Piesman J (2000). "Temporal changes in outer surface proteins A and C of the Lyme disease-associated spirochete, *Borrelia burgdorferi*, during the chain of infection in ticks and mice". *J Clin Microbiol* **38**(1):382-8.

Seltzer EG, Gerber MA, Cartter M L, et al. (2000). "Long-term outcomes of persons with Lyme disease". *JAMA* **283**(5):609-16.

Shadick NA, Phillips CB, Logigian E L, et al. (1994). "The long-term clinical outcomes of Lyme disease: a population-based retrospective cohort study". *Ann Intern Med* **121**(8):560-7.

Shapiro ED, Gerber MA, Holabird NB, et al. (1992). "A controlled trial of antimicrobial prophylaxis for Lyme disease after deer-tick bites". *N Engl J Med* **327**(25):1769-73.

Sjowall J, Ledel A, Ernerudh J, et al. (2012). "Doxycycline-mediated effects on persistent symptoms and systemic cytokine responses post-*neuroborreliosis*: a randomized, prospective, cross-over study". *BMC Infect Dis* **12**:186.

Skogman BH, Glimaker K, Nordwall M, *et al. (2012).* "Long-term clinical outcome after Lyme *neuroborreliosis* in childhood". *Pediatrics* **130**(2):262-9.

Smith K and Leyden JJ (2005). "Safety of doxycycline and minocycline: a systematic review". *Clin Ther* **27**(9):1329-42.

Sperling J, Middelveen M, Klein D, and Sperling F (2013). "Evolving perspectives on Lyme *borreliosis* in Canada". *Open Neurol J* **6**:94-103.

Stanek G and Reiter M (2011). "The expanding Lyme *borrelia* complex: clinical significance of genomic species?" *Clin Microbiol Infect* **17**(4):487-93.

Steere AC, Malawista SE, Newman JH, *et al.* (1980). "Antibiotic therapy in Lyme disease". *Ann Intern Med* **93**(1):1-8.

Steere AC, Hutchinson GJ, Rahn DW, *et al.* (1983a). "Treatment of the early manifestations of Lyme disease". *Ann Intern Med* **99**(1):22-6.

Steere AC, Bartenhagen NH, Craft JE, *et al.* (1983b). "The early clinical manifestations of Lyme disease". *Ann Intern Med* **99**(1):76-82.

Steere AC, Green J, Schoen RT, *et al.* (1985). "Successful parenteral penicillin therapy of established Lyme arthritis". *N Engl J Med* **312**(14):869-74.

Stricker RB (2007). "Counterpoint: long-term antibiotic therapy improves persistent symptoms associated with Lyme disease". *Clin Infect Dis* **45**(2):149-57.

Stricker RB, Johnson L (2007). "Lyme disease: a turning point". *Expert Rev Anti Infect Ther* **5**(5):759-62 [Taylor & Francis Online].

Stricker RB, Green CL, Savely VR, *et al.* (2010). "Safety of intravenous antibiotic therapy in patients referred for treatment of neurologic Lyme disease". *Minerva Med* **101**(1):1-7.

Strle F, Ruzic E, Cimperman J. (1992). "*Erythema migrans*: comparison of treatment with azithromycin, doxycycline and phenoxymethylpenicillin". *J Antimicrob Chemother* **30**(4):543-50.

Strle F, Preac-Mursic V, Cimperman J, *et al.* (1993). "Azithromycin versus doxycycline for treatment of *erythema migrans*: clinical and microbiological findings". *Infection* **21**(2):83-8.

Strle F, Maraspin V, Lotric-Furlan S, et al. (1996). "Azithromycin and doxycycline for treatment of *Borrelia* culture-positive *erythema migrans*" **24**(1):64-8.

Swanson SJ, Neitzel D, Reed KD, and Belongia EA (2006). "Coinfections acquired from *ixodes* ticks". *Clin Microbiol Rev* **19**(4):708-27.

Szczepanski A and Benach JL (1991). "Lyme *borreliosis*: host responses to *Borrelia burgdorferi*". *Microbiol Rev* **55**(1):21-34.

Tang S, Calkins H, and Petri M. (2004). "Neurally mediated hypotension in systemic lupus *erythematosus* patients with fibromyalgia". *Rheumatology* **43**(5):609-14.

Thaisetthawatkul P and Logigian EL (2002). "Peripheral nervous system manifestations of Lyme *borreliosis*". *J Clin Neuromuscul Dis* **3**(4):165-71.

Thompson GR 3rd, Lunetta JM, Johnson SM, *et al.*(2011). "Early treatment with fluconazole may abrogate the development of IgG antibodies in coccidioidomycosis". *Clin Infect Dis* **1**;**53**(6):e20-4.

Vazquez M, Sparrow SS, andShapiro ED. (2003). "Long-term neuropsychologic and health outcomes of children with facial nerve palsy attributable to Lyme disease". *Pediatrics* **112**(2):e93-7.

Volkman D. (2008). "Chemoprophylaxis against Lyme disease". *Lancet Infect Dis* **8**(3):145.

Warshafsky S, Nowakowski J, Nadelman R B, *et al.* (1996). "Efficacy of antibiotic prophylaxis for prevention of Lyme disease". *J Gen Intern Med* **11**(6):329-33.

Warshafsky S, Lee DH, Francois L K, *et al.* (2010). "Efficacy of antibiotic prophylaxis for the prevention of Lyme disease: an updated systematic review and meta-analysis". *J Antimicrob Chemother* **65**(6):1137-44.

Weber K, Neubert U, and Thurmayr R (1987). "Antibiotic therapy in early *erythema migrans* disease and related disorders". *Zentralbl Bakteriol Mikrobiol Hyg A* **263**(3):377-88.

Weber K, Wilske B; Preac-Mursic V, and Thurmayr R. (1993). "Azithromycin versus penicillin V for the treatment of early Lyme *borreliosis*". *Infection* **21**(6):367-72.

Weder B, Wiedersheim P, Matter L, *et al.* (1987). "Chronic progressive neurological involvement in *Borrelia burgdorferi* infection". *J Neurol* **234**(1):40-3.

Wormser GP, Ramanathan R, Nowakowski J, *et al.*(2003). "Duration of antibiotic therapy for early Lyme disease: a randomized, double-blind, placebo-controlled trial". *Ann Intern Med* **138**(9):697-704.

Wormser GP, Dattwyler RJ, Shapiro E D, *et al.* (2006). "The clinical assessment, treatment, and prevention of Lyme disease, human granulocytic *anaplasmosis* and *babesiosis*: clinical practice guidelines by the Infectious Diseases Society of America". *Clin Infect Dis* **43**(9):1089-134.

Zeidner NS, Brandt KS, Dadey E, *et al.* (2004). "Sustained-release formulation of doxycycline hyclate for prophylaxis of tick bite infection in a murine model of Lyme *borreliosis*". *Antimicrob Agents Chemother* **48**(7):2697-9.

Zeidner NS, Massung RF, Dolan MC, *et al.* (2008). "A sustained-release formulation of doxycycline hyclate (Atridox) prevents simultaneous infection of *Anaplasma phagocytophilum* and *Borrelia burgdorferi* transmitted by tick bite". *J Med Microbiol* **57**(Pt 4):463-8.

Zhang JR, Hardham JM, Barbour AG, and Norris SJ (1997). "Antigenic variation in Lyme disease *borreliae* by promiscuous recombination of VMP-like sequence cassettes". *Cell* **89**(2):275-85.

Lyme disease

Illustrations

1.1 The Old Lyme historic district. 15
1.2 View of Lyme's countryside. 16
1.3 Railroad Bridge over the Four-Mile River's mouth. 16
1.4 Official seal of Old Lyme, Connecticut. 17

2.1 Dr. W. Burgdorfer (in 1978) who first described the bacterium Borrelia burgdorferi. 19
2.2 Borrelia bacteria, the principal causative agents of Lyme disease. 20
2.3 Live Borrelia burgdorferi sensu stricto stained with wheat germ agglutinin. 21
2.4 The adult deer tick, Ixodes scapularis, the primary vector of Lyme disease in Central
　　and Eastern U.S., but also elsewhere such as in Europe 25
2.5 Adult deer tick, Ixodes scapularis. 26
2.6 Developmental stages of Ixodes scapularis. 27
2.7 Further illustration of Ixodes scapularis developmental stages. 28
2.8 Life cycle diagram of the deer tick Ixodes scapularis 29
2.9 Circulation of pathogenic European genospecies of Borrelia burgdorferi sensu lato. 30

4.1 Pictorial illustrating the signs and symptoms of Lyme disease. 49
4.2 "Classic" erythema migrans rash characteristic of a Lyme infection. 50
4.3 Illustrating facial palsy. 55
4.4 Illustrating a swollen knee. 55

6.1 CDC&P guideline for Lyme disease serology. 76

9.1 Points at which interruption of Borrelia burgdorferi transmission in humans
　　can be achieved through vaccination. 102

9.2 Temporally changing the composition of tick saliva spit into the host's skin. 104

S10.1 Body parts inspection for attached ticks. 115
S10.2 Dog body parts inspection for attached ticks. 116
S10.3 Using landscaping to create a tick-safe zone in your yard. 117
S10.4 Procedure for removing attached ticks. 119

11.1 Estimated geographical distribution of lone star ticks. 122

S12.1.1 Contrasting a healthy brain with a severe Alzheimer's diseased brain. ... 128
S12.1.2 Alzheimer's disease; demystifying the disease and you can do about it. ... 129
S12.2.1 Fulminating acute disseminated encephalomyelitis. ... 130
S12.3.1 Depicting the meninges of the central nervous system. ... 131
S12.4.1 Photomicrograph of a demyelinating multiple sclerosis lesion. ... 133
S12.5.1 Bell's palsy. ... 134
S12.6.1 Amyotrophic lateral sclerosis. ... 136

13.1 An abnormal heart rhythm can indicate Lyme carditis ... 140

15.1 Geographical areas of the U.S. affected by Lyme disease. ... 153
15.2 Annual reported cases of Lyme disease (1991-2017). ... 154

17.1 Number of U.S. anaplasmosis cases reported to CDC&P (2000-2017). ... 172
17.2 Number of reported U.S. anaplasmosis by month of onset (2000-2017). ... 173
17.3 Geographical distribution in the U.S. of anaplasmosis ... 175
17.4 Estimated geographical distribution of the Lone Star tick in the U.S. ... 177

18.1 A reference manual for tick-borne diseases of the United States. ... 183
18.2 Geographical areas in the U.S. affected by Anaplasmosis ... 188
18.3 Geographical areas in the U.S. affected by Babesiosis ... 188
18.4 Geographical areas in the U.S. affected by Ehrlichiosis ... 189
18.5 Geographical areas in the U.S. affected by Lyme disease. ... 189
18.6 Geographical areas in the U.S. affected by spotted fever rickettsiosis
        (including Rocky Mountain spotted fever) ... 190
18.7 Geographical areas in the U.S. affected by tularemia ... 191
22.1 Pregnancy and Lyme disease ... 226

Tables

2.1 Ticks of the United States and their characteristics ... 23
2.2 Tick-associated diseases of the United States ... 34

4.1 Signs and symptoms of early-localized infection (3-30 days after tick bite) ... 51

Lyme disease

4.2 Signs and symptoms of early-disseminated Lyme infection (days or months after tick bite)............53
4.3 Signs and symptoms of late-disseminated Lyme infection..56
5.1 Diagnostic methodology: localized (early) Lyme disease 62
5.2 Diagnostic methodology: early-disseminated Lyme disease 63
5.3 Diagnostic methodology: late-disseminated Lyme disease 69
5.4 Differentiating Lyme from confounding diseases...71
6.1 Recommended Lyme disease post-exposure prophylaxis 79
7.1 Treatment regimen for localized Lyme disease 85
7.2 Regimens for people with antibiotic resistance and neurological or cardiac illnesses 86
8.1 Botanicals for Lyme treatment and claimed characteristics..94
S9.1 Prospective Lyme vaccine research and development...107
S9.2 Positive and negative characteristics of the OspA vaccine..108
10.1 Tick killers and repellents... 111
14.1 State of knowledge of the cytomegalovirus.. 143
14.2 Pathogens of the brain 144
14.3 Eight types of Herpes viruses infect the brain.. 148
15.1 Lyme disease surveillance by State 154
15.2 Reported cases of Lyme disease by age groups 156
15.3 Lyme disease rates among reported (confirmed and probable) by
age group and sex (2017 157
17.1 Diseases transmitted by the black-legged tick Ixodes scapularis... 170
17.2 Number of anaplasmosis cases reported (2000-2017).. 171
17.3 Incidence rate of anaplasmosis by State (2017)... 174
18.1 Characteristics of tick-borne diseases of the United States 184

Sidebars

2.1 A brief look at tick-borne diseases of the United States...33
3.1 Chronic fatigue syndrome 42
3.2 Joints pain.. 44
3.3 Autoimmune diseases 45
3.4 Brain fog..46

6.1 Recommended Lyme disease post-exposure prophylaxis 79

7.1 Treatment regimen for localized Lyme disease. 85
7.2 Regimens for people with antibiotic resistance and neurological or cardiac illnesses. 86

8.1 Botanicals for Lyme treatment and claimed characteristics 94
S9.1 Prospective Lyme vaccine research and development 107
S9.2 Positive and negative characteristics of the OspA vaccine 108

10.1 Tick killers and repellents 111

14.1 State of knowledge of the cytomegalovirus 143
14.2 Pathogens of the brain. 144
14.3 Eight types of Herpes viruses infect the brain. 148

15.1 Lyme disease surveillance by State. 154
15.2 Reported cases of Lyme disease by age groups. 156
15.3 Lyme disease rates among reported (confirmed and probable) by
    age group and sex (2017). 157

17.1 Diseases transmitted by the black-legged tick Ixodes scapularis. 170
17.2 Number of anaplasmosis cases reported (2000-2017). 171
17.3 Incidence rate of anaplasmosis by State (2017). 174

18.1 Characteristics of tick-borne diseases of the United States. 184

Sidebars

2.1 A brief look at tick-borne diseases of the United States. 33

3.1 Chronic fatigue syndrome. 42
3.2 Joints pain. 44
3.3 Autoimmune diseases. 45
3.4 Brain fog. 46

6.1 Evidence-based guidelines for the management of LD patients 81
6.2 International Lyme and Associated Diseases Syndrome management and treatment guidelines. ... 83

9.1 Further considerations on Lyme vaccine research and development. 107

10.1 Measures for preventing tick bites. 114
10.2 Removing attached ticks. 118

12.1 Alzheimer's disease. 127
12.2 Acute disseminated encephalomyelitis. 129
12.3 Viral meningitis. 131
12.4 Multiple sclerosis. 132
12.5 Bell's palsy. 133
12.6 Amyotrophic lateral sclerosis 135

14.1 Overview of Herpes simplex viruses that infect the brain. 148
14.2 Familial and autoimmune encephalitis. 149

19.1 A sample clinical trial: novel diagnostics for early Lyme disease. 203

Glossary

A

Acaricides: Tick pesticides.

Acrodermatitis chronica atrophicans: A chronic skin disorder observed primarily in Europe among the elderly. It begins as a reddish-blue patch of discolored skin, often on the backs of the hands or feet. The lesion slowly atrophies over several weeks or months, with the skin becoming first thin and wrinkled and then, if untreated, completely dry and hairless.

Anaplasmosis: An infectious disease of ruminants varying from peracute to chronic. It is caused by Anaplasma species. It is characterized by anemia, icterus, and fever. Blood-feeding arthropods, chiefly ticks may serve as vectors.

Ankylosing spondylitis: Inflammation of one or more of the vertebrae resulting in a stiffening or fixation of a joint or joint structure as a result of a disease process with fibrous or bony union across the joint.

Anthropod: Resembling man in structure and form.

Antibody: Protective protein used to identify bacteria and viruses. It is evoked by an antigen. It is characterized by reacting specifically with the antigen in a demonstrable way. Antibody and antigen are each defined in terms of the other. It is now supposed that antibodies may exist naturally without being present as a result of the stimulus provided by the introduction of an antigen.

Antigen: Allergen; immunogen; any substance that, as a result of coming in contact with appropriate tissues, induces a state of sensitivity and/or resistance to infection or toxic substances after a latent period (8-14 days), It reacts in a demonstrable way with tissues and/or antibody of the sensitized subject.

Ataxia: Inability to coordinate the muscles in the execution of voluntary movement.

Athralgia: Severe pain in a joint, especially one not inflammatory in character.

Arthritis: The inflammation of the synovium of the joints. There are different manifestations: juvenile idiopathic; juvenile rheumatoid; osteoarthritis; psoriatic; rheumatoid arthritis.

Arthropod: A member of the phylum Arthropoda.

Autoimmune disease: The disease resulting from an immune reaction produced by an individual's leukocytes or antibodies acting on the subject's own tissues or extracellular proteins.

Autoimmunity: The condition in which one's own tissues are subject to deleterious effects of the immune system.

Axons: The single one among a nerve cell's processes that under normal conditions conducts nervous impulses from the cell body an its remaining cell processes (dendrites).

B

Babesiosis: A highly pathogenic disease caused by infection with a species of *Babesia,* the infection being transmitted by ticks. The disease is characterized by fever, malaise, listlessness, severe anemia, and hemoglobulinuria.

Baker's cyst: A collection of synovial fluid which has escaped from the knee joint or a bursa and formed a new synovial-lined sac in the popliteal space. Seen in degenerative or other joint disease.

Bell's palsy: Most common type of one-sided, facial palsy.

Bannwarth syndrome: A combination of lymphocytic meningitis and radiculopathy.

Bartonellosis: Disease caused by *Bartonella bacifillormis,* transmitted by the bite of the nocturnally biting sandfly.

Biofilm: A protein mesh a microbe secretes to shield it and its progeny from direct exposure to many antibiotics and other compounds that might otherwise harm the infection.

Blood-brain barrier: A highly selective, semi-permeable border that separates the circulating blood from the brain and extracellular fluid in the central nervous system.

Borrelial **lymphocytoma:** A purplish lump that develops on the ear lobe, nipple, or scrotum.

Brain fog: Build-up of toxins in the brain that causes memory issues.

C

C3: A protein marker.

Carcinoma: Any of the various types of malignant neoplasm derived from epithelial tissue in several sites, occurring more frequently in the skin and large intestine, the bronchi, stomach, prostate, breast and cervix.

Cardiomegaly: Enlargement of the heart that usually takes some time to develop and is often a result of a circulatory issue causing the heart to grow in an effort to compensate for dysfunction elsewhere.

Carditis: Acute inflammation of the heart.

Case fatality rate: Proportion of disease patients that reportedly died as a result of infection by the disease.

Cellulitis: Inflammation of cellular or connective tissue.

Colitis: Inflammation of the colon.

Congenital cytomegalic inclusion disease: Disease due to the presence of inclusion bodies within the cytoplasm and nuclei of enlarged cells of various organs of newborn infants dying with jaundice, hepatomegaly, splenomegaly, purpura, thrombocytopenia, and fever. Also occurs at all ages as a complication of other diseases in which immune mechanisms are severely depressed.

Cranial neuritis: Inflammation of cranial nerves. When due to Lyme, it most typically causes facial palsy, impairing blinking, smiling, and chewing in one or both sides of the face. It may also cause intermittent double vision.

Creutzfeldt-Jakob disease: Spastic pseudo-sclerosis with corticostriatal-spinal degeneration. A form of spongiform encephalopathy caused by a slow virus and characterized by dementia, myoclonus, ataxia, and other neurologic manifestations. Progresses rapidly to coma and death.

Crohn's disease: Regional enteritis.

Cytokines: Immune cells that generate inflammation to fight the infection.

Cytomegalovirus: A Lyme disease infection and cancer-causing agent. It is a member of the herpes family. It has the ability to remain alive, yet dormant, in the human body for the life of the human host. It rarely becomes active unless the immune system is weakened and rendered unable to hold the virus in check.

D

DEET: A tick-repellent.
Depression: Sinking below the normal functional level constituting a clinically-discernible condition.
Dermatome: Instrument for cutting thin slices of skin for grafting or excising small lesions.

E

Empyema: Suppuration. Pus in a body cavity.
Encephalitis: Brain inflammation occurring when a virus, vaccine or something else triggers it. There are several forms: eastern equine encephalitis, which affects mainly young children and people older than 55. In children younger than 1 year, it can cause severe symptoms and permanent nerve or brain damage; herpes labial encephalitis, which is caused by a herpetic infection of the lips from the herpes virus type 1; La Crosse encephalitis, which is caused by the La Crosse virus (also called California virus). It accounts for most cases in children. Many cases are mild and undiagnosed; Sassanian virus, which : It usually causes mild or no symptoms. The virus is similar to the one that causes tick-borne encephalitis. However, the infection can also cause severe encephalitis with headache, vomiting, seizures, loss of coordination, speech problems, or coma. The vaccine that is effective against tick-borne encephalitis is not effective against the Sassanian virus; St. Louis, which is more likely to affect the brain in older people. Epidemics once occurred about every 10 years but are now rare; tick-borne encephalitis, which is caused by a tick bite; western equine, which, for unknown reasons, has largely disappeared since 1988. It can affect all age groups but is more severe and more likely to affect the brain in children younger than 1.
Encephalomyelitis: Encephalitis involving the spinal cord.
Encephalopathy: Any disease of the brain.
Endocarditis: Inflammation of the endocardium or lining membrane of the heart.
Endotoxin: Bacterial toxin not freely liberated into the surrounding medium.
Enteroviruses: Common stomach viruses.
Entomology: The science concerned with insects, especially as it affects man. Medical entomology he study of the interactions between animal and human disease agents and their transmitting arthropod vectors, particularly ticks, fleas, and mosquitoes.
Enzyme: Protein secreted by cells. Acts as a catalyst to induce chemical changes in other substances, itself apparently remaining unchanged in the process.
Epstein-Barr virus: A herpetovirus found in cell cultures of Burkitt's lymphoma. Also found in infection mononucleosis.
Erythema migrans: Rash at the site of a tick bite, often but not always near skin folds, such as the armpit, groin, the back of the knee, on the trunk, under clothing straps, or in children's hair, ear, or neck. Most people who get infected do not remember seeing a tick or the bite.
Evidence-based medicine: The integration of clinical research, clinical expertise, and patient values and preferences.

F

Facial palsy: Loss of muscle tone or droop on one or both sides of the face.
False negative: The situation occurs when a test produces results indicating that a disease is not present when, in reality, it is.
Fungus: General term used to encompass the diverse morphological forms of yeasts and molds.

G

Genomic testing: Genomic testing is based on genomic information from the patient's blood, urine, mucus, and stool to derive important DNA and RNA information without having to synthetically amplify the genes for detection as required by the Polymerase Chain Reaction method. Not only allowing for a better diagnosis, it also paves the way for a personalized, comprehensive treatment plan and quantifies the amount of infections present.

Glial cells: Pertaining to glia or neuroglia.

H

Heart block: The cardiac condition where the four chambers of the heart are not communicating correctly in order to manage the valves and the emptying and filling of the sections of the heart.

Hodgkin's disease: Chronic enlargement of the lymph nodes.

I

Idiopathic: A condition with no defined cause.

Incidence: Number of cases for every million people.

Infectious disease reservoirs; Places or populations that harbor disease-causing pathogens.

Immunoglobulins: Proteins that work as antibodies:

Ig: Immunoglobulin. One of a class of structurally-related proteins consisting of two pairs of polypeptide chains, one pair of light and the other pair of heavy chains, all four linked by disulfide bonds. Classified in relative amounts present in human normal serum as IgG (80%; the biggest antibody in the human circulatory system), IgA)15%), IgM (5-10%), IgD (less than 0.1%), and IgE (less than 0.01%). All of these classes are homogeneous and susceptible to amino-acid sequence analysis.

Infectious disease reservoirs: Places or populations that harbor disease-causing pathogens.

Intrathecal space: Ensheathed; within a sheath.

IR3535: A derivative of the amino-acid alanine and a tick-repellent.

J

Juvenile: Onset of disease in children aged 16 or less.

K

Kaposi sarcoma: A multifocal mailgnant or benign neoplasm of primitive vasoformative tissue, occurring in the skin and sometimes in lymph nodes or viscera.

L

Leukoplakia: A disturbance of keratinization of a mucous membrane variously present as small or as extensive leathery plaques opalescent patches, occasionally ulcerated.

Lupus: Generally used to depict erosion (as if gnawed) of the skin.

Lyme arthritis: Usually affects only one or a few joints, often a knee or possibly the hip, other large joints, or the temporomandibular joint. There is usually large joint effusion and swelling, mild or moderate pain. Without treatment, swelling and pain typically resolve over time but periodically return.

Baker cysts may form and rupture. In some cases, joint erosion occurs. Chronic neurologic symptoms occur in up to 5% of untreated people. A peripheral neuropathy or polyneuropathy may develop, causing abnormal sensations such as numbness, tingling or burning starting at the feet or hands and over time possibly moving up the limbs. A test may show reduced sensation of vibrations in the feet. An affected person may feel as if wearing a stocking or glove without actually doing so.

Lyme carditis: Heart complications due to Lyme disease. Symptoms may include heart palpitations, dizziness, fainting, shortness of breath, and chest pain. Other symptoms may also be present, such as EM rash, joint aches, facial palsy, headaches, or radicular pain. In some people, however, carditis may be the first manifestation of LD. It may adversely impact the heart's electrical conduction system, causing atrioventricular block that often manifests as heart rhythms that alternate within minutes between abnormally slow and abnormally fast. In 10%-15% of people. Lyme causes myocardial complications such as cardiomegaly, left ventricular dysfunction, or congestive heart failure.

Lyme disease - early-localized: When the infection has not yet spread throughout the body. Only the site where the infection has first come into contact with the skin is affected.

Lyme disease - early-disseminated: Within days to weeks after the onset of local infection, the *Borrelia* bacteria may spread through the lymphatic system or bloodstream.

Lyme disease – late-disseminated (or "late persistent infection") - After several months, untreated or inadequately treated people may go on to develop chronic symptoms that affect many parts of the body, including the joints, nerves, brain, eyes, and heart.

Lyme encephalopathy: A neurologic syndrome associated with subtle memory and cognitive difficulties, insomnia, a general sense of feeling unwell, and changes in personality. However, problems such as depression and fibromyalgia are as common in people with Lyme disease as in the general population.

Lyme *neuroborreliosis*: Neurological problems caused by Lyme disease. Involves some combination of cranial neuritis, lymphocytic meningitis, radiculopathy and/or mononeuritis multiplex.

Lyme radiculopathy: Inflammation of spinal nerve roots that often causes pain and less often weakness, numbness, or altered sensation in the areas of the body served by nerves connected to the affected roots, for example, the limb(s) or part(s) of the trunk. The pain is often described as unlike any other previously felt, excruciating, migrating, worse at night, rarely symmetrical, and often accompanied by extreme sleep disturbance.

Lymphocyte: Lymph cell; lymphoid cell; white blood cell.

Lyme lymphocytic meningitis: It causes characteristic changes in the cerebrospinal fluid. For several weeks, it may be accompanied by variable headache and usually, but less commonly, mild meningitis signs such as inability to flex the neck fully and intolerance to bright lights, but typically no or only very low fever. In children, partial loss of vision may also occur.

Lymphocytic pleocythosis: Densities of lymphocytes (infection-fighting cells) and proteins in the cerebrospinal fluid. They typically rise to characteristically abnormal levels while glucose remains normal.

Lymphoma: Ordinarily malignant neoplasm of lymph and reticuloendothelial tissue, which presents as apparently circumscribed solid tumor composed of cells that appear primitive or resemble lymphocytes, plasma cells, or histiocytes.

Lysate: The material (cellular debris and fluid) produced by lysis.

Lyse: To break-up, to disintegrate, for example, as applied to cells. To effect lysis.

Lysin: A specific complement-fixing antibody that acts destructively on cells and tissues.

Lysis: The gradual subsidence of the symptoms of an acute disease, a form of the curative process, distinguished from crisis. Also, the destruction of red blood cells, bacteria, and other antigens (a specific lysin). According to the form of antigen destroyed, the process is called hemolysis, nephrolysis, bacteriolysis, and so on.

M

Malaria: Disease caused by the presence of the protozoan *Plasmodium*. The agent is transmitted to humans by the bite of an infected female mosquito of the genus *Anophele* that previously sucked the blood of a person infected with it.
Meninges: The membranes covering the brain and spinal cord.
Meningitis: Inflammation of the meninges.
Metabolomics: A type of science that can be used to identify and measure types and amounts of chemicals the body produces during illness. Each type of infection or stage of infection has a different metabolic "fingerprint" that makes it unique.
Microglia: Small neuroglial cells of mesodermal origin which may become phagocytic, hence are considered elements of the reticuloendothelial system.
Mononeuritis multiplex: Inflammation causing similar symptoms in one or more unrelated peripheral nerves. Rarely, early *neuroborreliosis* may involve inflammation of the brain or spinal cord with symptoms such as confusion, abnormal gait, ocular movements, slurred speech, impaired movement, impaired motor planning, or shaking (a Parkinsonian syndrome).
Mononucleosis: The presence of abnormally large numbers of mononuclear leukocytes in the circulating blood, especially with reference to forms that are not normal.
Myelin sheath: Cells that cover and insulate nerves.
Myocarditis: Inflammation of the myocardium or muscular walls of the heart.
Myopericarditis: Inflammation of the muscular wall of the heart and of the enveloping pericardium.

N

Necrosis: Sinking dark-blue patches of dead skin.
Neuritis: Inflammation of a nerve.
Neuron: Nerve cell.
Neurotransmitters: Nerve messengers (examples are dopamine, octopamine).

O

Oil of lemon eucalyptus: A natural tick-repellent.
OspA (outer surface protein A): An immunogenic lipoprotein whose expression is abundant on *in vitro*-cultured spirochetes and spirochetes within the tick midgut. Immunization with OspA provides cross-protection of mice challenged with the North American isolates of *Borrelia burgdorferi*.
OspC: Antibodies from this outer surface protein kill any of the bacteria that have not been killed by the OspA antibodies.
Otitis: Inflammation of the ear.

P

Palsy: Partial paralysis or paresis. Bell's palsy is a facial palsy.
Pathogen: Any virus, microorganism or other substance causing disease.

Pericarditis: Inflammation of the pericardium.
Permethrin: An odorless substance that is safe for humans but highly toxic to ticks. It is recommended by the CDC & P as a preventive measure against tick bites.
Picardin: A tick-repellent.
Plasmapheresis: A procedure to filter the blood.
Plasmid: Extrachromosomal element.
PMD: A tick-repellent.
Polymerase: Any enzyme catalyzing a polymerization.
Polymorphonuclear: Having nuclei of varied forms, denoting a variety of leukocytes.
Prion: Abnormal protein.

R

Resisters: Organisms remaining despite antibiotic therapy.

S

Sarcoma: A connective tissue neoplasm, usually highly malignant, formed by proliferation of mesodermal cells.
Sclerae: Essentially plaques or lesions.
Seropositive: Bacterium present in the blood.
Somatoform: Combining forms denoting the body; bodily.
Spirochaeta: A genus of motile bacteria containing presumably Gram-negative flexible undulating, spiral-shaped rods which may or may not possess flagelliform, tapering ends.
Spirochete: A vernacular term used to refer to any member of the genus *Spirochaeta*. It is shaped like a corkscrew.
Synaptoblastic: Enhancing the formation of synapses.
Synaptoclastic: Depressing the formation of synapses.
Syphillis: It is caused by *Triponema pallidium* (a spirochete type of bacterium). It can live in the body for decades, eventually infecting the brain and causing dementia.

V

Vaccine - oral bait: Aimed at preventing mice from becoming infected, thereby interrupting the transmission cycle.
Vaccine - rice-based: Rice plants contain vaccine elements that could eventually be fed to rodent populations, thus blocking the transmission cycle of the disease from rodents to ticks to people.
Vaccinia virus: A mouse-targeted vaccine using this virus can provide full protection with a single dose.

Z

Zoonosis: Infection or infestation shared in nature by men and lower vertebrate animals.
Zoonotic disease: Disease transmitted from lower vertebrate animals to men.

Abbreviations

A

AAN: American Academy of Neurology
AB: Antibody
AB3CE: *Anaplasmosis, Babesiosis, Bartonellosis, Borreliosis,* Colorado tick fever, *Ehrlichiosis*
AC2EMPQ: *Anaplasmosis,* Chlamydia pneumonia, Coltivirus disease, *Ehrlichiosis,* Mycoplasma pneumonia, Powassan virus disease, Q-fever
ACA: *Adenomatitis Chronica Atrophicans*
AD: Alzheimer's Disease
ADA: *Acrodermatitis Chronica Atrophicans*
ADE: Acute Demyelinating Encephalomyelitis
ADEM: Acute Disseminated Encephalomyelitis
ADM: Advanced Molecular Detection
AHRQ: (U.S.) Agency for Healthcare Research Quality
ALS: Amyotrophic Lateral Sclerosis (aka Lou Gehrig disease)
AM: Aseptic Meningitis
AME: Acute Demyelinating Encephalomyelitis
ASTPHD: Association of State and Territorial Public Health Directors
AV: Atrio-Ventricular

B

Bb: *Borrelia burgdorferi*
Bbsl: *Borrelia burgdorferi sensu lato* complex
Bbss: *Borrelia burgdorferi sensu stricto*
BBB: Blood-Brain Barrier
BL: *Borrelial* Lymphocytoma
BM: Bacterial Meningitis
BP: Ball's Palsy
BPB: Brain Protective Barriers
BS: Bannworth Syndrome
BUN: Blood Urea Nitrogen

C

C2MPQ: Chlamydia pneumonia, Coltivirus disease, Mycoplasma pneumonia, Powassan virus, Q-fever disease
CAES: Connecticut Agricultural Experiment Station
CAS: Coronary Artery Syndrome
CD: Crohn's Disease
CDC&P: (U.S.) Centers for Disease Control & Prevention
CDMRP: Congressionally-Directed Medical Research Program
CFS: Chronic Fatigue Syndrome
CHF: Congestive Heart Failure
CIDP: Chronic Inflammatory Demyelinating Polyneuropathy
CJD: Creutzfeldt-Jakob Disease
CLIA: Clinical Laboratory Improvement Amendment compliance
CLD: Chronic Lyme Disease
CLDC: Chronic Lyme Disease Complex
CLD-PT: Previously treated CLD
CLDS: Chronic Lyme Disease Syndrome
CLD-U: Untreated CLD
CMV: Cytomegalovirus
CN: Cranial Neuritis
CNS: Central Nervous System
CPG: Clinical Practice Guidelines
CSF: Cerebrospinal Fluid
CSTE: Council of State and Territorial Epidemiologists

D

DHHS: (U.S.) Department of Health & Human Services
DRG: Dose-Response Gradient
DNA: Deoxyribonucleic Acid

E

EBV: Epstein-Barr Virus
EIA: Enzyme ImmunoAssay
EKG: Electrocardiogram
EL: Encephalitis Lethargica
ELCID: Epidemiology and Laboratory Capacity for Infectious Diseases

ELISA: Enzyme-Linked ImmunoSorbent Assay
EM: Enteroviral Meningitis
EM: *Erythema Migrans*
EPA: (U.S.) Environmental Protection Agency.
ES: Encephalopathy Spongiform

F
FDA: (U.S.) Food & Drug Association
FISH: Fluorescent *In Situ* Hybridization
FNN: Foundation for Neural Neuropathy
FNP: Facial Nerve Palsy
FP: Facial Palsy

G
GBBS: Gain-Boujadoux-Bannwarth Syndrome
GBS: Guillain-Barre Syndrome
GD: Graves' Disease
GRADE: Grading of Recommendations, Assessment, Development, and Evaluation
GSK: Glaxo-Smith-Kline (a pharmaceutical company)

H
HHV: Human Herpes Virus
HHSV: Human Herpes Simplex Virus
HIV: Human Immunodeficiency Virus
HLA: Hypothalamus-Pituitary-Adrenal (axis)
HSV: Herpes Simplex Virus
HT: Hashimoto's Thyroiditis

I
IBA: ImmunoBlot Assay
IBD: Inflammatory Bowel Disease
IDSA: (U.S.) Infectious Diseases Society of America
IFA: Indirect Fluorescent Antibody
IFA: Immunofluorescent Assay
Ig: Immunoglobulin
ILADS: International Lyme and Associated Diseases Society
IOM: (U.S.) Institute of Medicine (now: National Academy of Medicine)
IV: Intra-Venous

K
KSHV: Kaposi Sarcoma-associated Herpes Virus

L
LA: Lyme Arthritis
LC: Lyme carditis
LCM: Lymphocytic Choriomeningitis
LD: Lyme Disease
LDC: Lyme Disease Complex
LE: Lyme Encephalopathy
LDFP: Lyme Disease Facial Palsy
LLM: Lyme Lymphocytic Meningitis
LM: Lymphocytic Meningitis
LMR: Lymphocytic Meningo-Radiculopathy
LN: Lymphocytic neuritis
LNB: Lyme *Neuroborreliosis*
LPC: Lymphocytic Pleocytosis
LR: Lyme Radiculopathy

M
MDEM: Multiphasic Disseminating Encephalomyelitis
ME/CFS: Myalgic Encephalomyelitis/Chronic Fatigue Syndrome
MEM: Multiple *Erythema Migrans*
MG: Myasthena Gravis
MND: Motor Neuron Disease
MNM: Mononeuritis Multiplex
MRI: Magnetic resonance Imaging
MS: Multiple Sclerosis

N
NAM: (U.S.) National Academy of Medicine
NB: *Neuroborreliosis*
NCCLS: National Committee for Clinical Laboratory Standards
NCEZID: National Center for Emerging and Zoonotic Infectious Diseases
NDD: Neurodegenerative Disease
NGC: National Guidelines Clearinghouse
NGS: Next Generation Sequencing
NIH: (U.S.) National Institutes of Health
NIANS: (U.S.) National Institute for Allergy &

Infectious Diseases
NINDS: National Institute of Neurological Diseases & Stroke
NLTBD: Non-Lyme Tick-Borne Disease
NNDSS: National Notifiable Disease Surveillance System
NNS: Neuropathy Network Support
NSAID: Non-Steroidal Anti-Inflammatory Drug

O

OLE: Oil of Lemon Eucalyptus
Os: Outer surface protein
OTC: Over-The-Counter

P

PCR: Polymerase Chain Reaction
PD: Parkinson's Disease
PEM: Post-Exertion Malaise
PHIL: Public Health Image Laboratory
PHS: (U.S.) Public Health Service
PLD: Post-Lyme Disease
PLDS: Post-Treatment Lyme Disease Syndrome
PMC: Pasteur-Merriment-Connaught (a pharmaceutical company)
PMD: Para-Methane-Diol
PML: Progressive Multifocal Electroencephalography
PLN: Polymorphonuclear Leukocytes
PNL: Polymorphonuclear Leukocytes
PNS: Peripheral Nervous System

R

RA: Rheumatoid Arthritis
RBC: Red Blood Cells
RCT: Randomized Clinical Trials
RDEM: Recurrent Disseminated Encephalomyelitis
RMSF: Rocky Mountain Spotted Fever
RNA: Ribonucleic Acid
RS: Reiter's Syndrome
RHS: Ramsey-Hunt Syndrome

S

SE: Spongiform Encephalitis
SEID: Systemic Exertion Intolerance Disease
SFD: Spotted Fever Rickettsiosis
SKB: Smith-Kline-Beech an (a pharmaceutical company; formerly GSK)
SLE: Systemic Lupus Erythematosus
SNCSDLD: Second National Conference on Serologic Diagnosis of Lyme Disease
SPECT: Single Photon Emission Computed Tomography
SS: Sjogren's Syndrome
SSP: Subacute Sclerosis Pan encephalitis
STARI: Southern Tick-Associated Rash Illness

T

TCM: Traditional Chines Medicine
TBD: Tick-Borne Disease
TBDWG: Tick-Borne Disease Working Group
TBRF: Tick-Borne Relapsing Fever
TENS: Transcutaneous Electrical Nerve Stimulation
T1DM: Type 1 Diabetes Mellitus
TMJ: Temporo Mandibular Joint

U

UC: Ulcerative Colitis

V

VM: Viral Meningitis

W

WGS: Whole Genome Sequencing
WNA: Western Neuropathy Association

Subject Index

A

Acute respiratory syndrome 178
Acrodermatitis chronica atrophicans 56-59, 68, 112
Adenopathy, lymph 185
Alzheimer's disease 12, 31, 125, 126, 144-148
Alzheimer's disease, early-onset 127
Alzheimer's disease, late-onset 127
Amblyoma, americanum (see also lone star tick) 33, 34, 121, 123, 176
Amblyoma, maculatum 24, 33, 34
Amphilobacter (see Guillain-Barre syndrome)
Amyotrophic lateral sclerosis 12, 126, 127, 135, 217-219
Anaplasma phagocytophilum 23, 109, 171, 172, 180
Anaplasmosis (*Anaplasma phagocytophylum*) 12, 23, 33-35, 37, 81, 112, 161, 170, 173-175, 177, 180, 188, 190
Ankylosing spondilitis 17
Antibiotic resistance 216
Antibiotics 40, 41, 51, 57, 62, 63, 66-68, 70, 74, 75, 77, 78, 80-93, 97, 112, 113, 119, 123, 124, 126, 127, 139, 140, 142, 161-164, 176, 180, 184-187, 200, 201, 206, 207, 213, 215, 216, 220, 232, 234, 237, 240
Antibiotics, prophylaxis 74
Antibodies 11, 39, 66
Antimicrobials, prophylaxis 81
Antivirals 75
Arthralgia 9, 21, 176, 178, 180, 181, 184, 185, 187, 236, 240
Arthritis (see also joints inflammation and pain) 44, 54-56, 58, 95, 96, 112, 139, 161, 198, 216, 220-222, 231, 233, 239
Arthritis, juvenile idiopathic 7
Arthritis, juvenile rheumatoid 7, 32
Arthritis, osteo 17, 44
Arthritis, psoriatic 17
Arthritis, rheumatoid 11, 38, 44, 45, 69, 73, 113
Arthritis, septic 44
Arthropod vector 20
Assay, enzyme immuno 75, 78, 80, 164
Assay, enzyme immunofluorescent 164
Assay, enzyme-liked immunosorbent assay (ELISA) 11, 60, 63, 66, 72, 73, 76, 77, 125, 176, 229, 232, 239
Assay, immunofluorescence 75, 77, 80, 197
Assay, indirect fluorescent antibody 63
Assay, Western (or striped) immunoblot 11, 60, 63, 64, 66, 72, 73, 75-78, 80, 81, 126, 164, 229, 232, 239
Ataxia, myelitis 125, 126
Autoimmune diseases 10, 38, 41, 44, 45, 66, 69, 73, 112, 130, 146, 149, 203
Autoimmunity 86

B

Babesia 12, 31, 39, 95, 145, 148, 161, 178-180, 184
Babesia, microtii 23, 33, 35, 109, 170, 178
Babesiosis 12, 23, 33, 35, 37, 81, 112, 170, 178, 180, 182, 187, 188, 190
Bacteria 146
Baker's cyst 55, 57, 58, 112
Bannwarth syndrome 53
Bartonella 12, 146, 148, 161
Bartonellosis 37
Biofilm 9, 13, 31, 33, 39, 94, 95, 98, 216, 221, 240
Biomarkers, metabolic 197
Biosignatures 197
Bipolar disorder 56, 221
Blood-brain barrier 9, 31, 32, 38, 39, 87, 126, 146
Borrelia 7-10, 18, 32, 37, 39, 40, 52, 59, 63, 65, 67, 69, 72, 86, 95, 96, 100, 101, 124, 146, 148, *220*
Borrelia afzelli 8, 27, 59, 68
Borrelia bavariansis 59
Borrelia burgdorferi 7-10, 13, 18-20, 22, 23, 25, 26, 28, 30, 32, 35, 36, 38, 40, 41, 50, 60, 63, 68, 77-79, 81, 86, 87, 89, 92, 93, 99-103, 105, 106, 108, 121, 122, 124, 126, 138, 145, 160, 162, 164, 170, 172, 180, 185, 196, 197, 216, 217, 220-222, 229-231, 236, 237, 240

Borrelia burgdorferi sensu lato 59, 230, 235, 237, 240
Borrelia burgdorferi sensu scricto 59
Borrelia garinii 8, 27, 59, 68
Borrelia hermsii (tick-borne relapsing fever) 24
Borrelia, lonestari 122
Borrelia, mayonii 19, 23, 25, 33, 35, 168, 170
Borrelia, miyamotoi 23, 33, 35, 112, 161, 175, 176, 179, 180, 182, 184, 185
Borrelia, turicatae (tick-borne relapsing fever) 24, 187
Borreliosis 7, 18, 32, 37, 64, 75, 80, 138, 180, 187, 199, 231, 239
Borreliosis, congenital 227
Botanicals 94
Brain fog 38, 41, 46, 47, 56, 58, 61, 94, 97, 209
Burning mouth syndrome 208
Bursitis 44

C

Cancers 41, 112, 146, 150, 218, 221
Carditis 54, 220, 222
Carditis, endo 138
Carditis, myo 138
Carditis, myoperi 138
Carditis, peri 138
Cardiomegaly 53, 54, 58, 138
Cell, persister 13
Cell, stem 13
Cellulitis 70, 71, 73, 113
Cerebrospinal fluid 52, 65, 67-69, 73, 90, 132, 179, 236
Chemokines 147, 200
Chlamydia (see Reiter's syndrome)
Chondromalacia patellae 44
Chorea 12, 125, 126
Clinical trials 126
Coagulation, disseminated intravascular 178
Colloid silver, therapy
Colitis, ulcerative 46
Coltivirus disease 37
Congestive heart failure 53, 58
Controversies and challenges 229
Corticosteroids 75

Coronary artery syndrome 71, 73, 113
Crohn's disease 11, 46, 69, 73, 113
Crutzfeldt-Jakob disease 69
Cytokines 37, 64, 67, 147, 197, 200
Cytomegalovirus 12, 61, 69, 143, 146, 147, 149
Cytopenia, trans 176

D

Delusional behavior 58, 112
Delusions, somatoform 56, 58, 112
Dementia 145
Depersonalization 56
Depression 58. 60, 112, 236, 240
Derealization syndrome 56
Dermacentor, andersonii 24, 33, 34, 109
Dermacentor, occidentalis 34, 109,
Dermacentor, variabilis 34, 109
Dermatitis, atrophic chronica 18
Detoxification 93, 94, 96-98
Diabetes mellitus, type 1 45, 135
Diverticulitis 71, 73, 113
Duchesne muscular dystrophy 218

E

Ehrlichia 12, 148
Ehrlichia, chaffensis 23, 170, 176-178, 180, 185
Ehrlichia, equi 171
Ehrlichia, ewingii 23, 170, 176, 178, 180, 185
Ehrlichia, muris eauclairensis 23, 170, 176, 178, 180, 185
Ehrlichia, phagocytophilum 171
Ehrlichiosis (human granulocytic) 33-35, 37, 81, 161, 170, 173, 176, 177, 180, 182, 185, 189, 190
Electrocardiogram 139-141
Electromyography 90, 92
ELISA (see Assays)
Encephalitis 69, 144, 146, 148, 149, 199, 221
Encephalitis, autoimmune 149, 151
Encephalitis, Chikungaya virus 151
Encephalitis, eastern equine 150
Encephalitis, familial 149
Encephalitis, Japanese virus 151
Encephalitis, La Crosse 150
Encephalitis, lethargica 145

Encephalitis, meningo 179, 181
Encephalitis, post-infectious 146
Encephalitis, Sassanian virus 150
Encephalitis, spread by mosquitos 151
Encephalitis, spread by ticks
Encephalitis, St Louis 150
Encephalitis, Venezuelan equine virus 151
Encephalitis, West Nile virus 150
Encephalitis, western equine 150
Encephalitis, Zika virus 151
Encephalomyelitis 125
Encephalomyelitis, acute disseminated 12, 125, 126, 129, 130, 131
Encephalomyelitis, acute demyelinated 130
Encephalomyelitis, chronic 6
Encephalomyelitis, chronic progressive 58, 112, 113
Encephalomyelitis, multiphasic 131
Encephalomyelitis, myalgic 42, 198
Encephalomyelitis, recurrent disseminated 131
Encephalopathy 10, 22, 58, 112, 216, 220
Encephalopathy, leuko progressive multifocal 151
Encephalopathy, spongiform 146
Endotoxins 38
Epidemiology 152
Epilepsy 217
Epstein-Barre virus 39, 42, 61, 148
Erythema migrans 121, 123, 125, 126, 139, 163, 185, 200, 202, 203, 215, 230, 232, 235, 239, 240
Evidence-based medicine 82
Exertion, systemic intolerance disease 42

F
Fatigue 41, 51, 57, 70, 71, 86-89, 91, 112, 121-124, 139, 143, 164, 176, 178, 180, 185, 236, 240
Fatigue, chronic syndrome 11, 31, 37, 42, 43, 113, 108, 236, 240
Fever, Colorado tick 20, 24, 33, 37, 112, 185
Fever, hard tick relapsing 175, 180
Fever, Q 37, 42
Fever, relapsing 20, 23
Fever, Rocky Mountain spotted (*Rickettsia rickettsiosis*) 7, 20, 24, 34, 35, 39, 78, 81, 112, 177, 182, 186, 187, 190
Fever, spotted *rickettsiosis* 182, 190
Fever, tick-borne relapsing (*Borrelia hermsii, Borrelia turicatae*) 34, 35, 112, 187
Fibromyalgia 11, 31, 56-58, 66, 69, 73, 112, 113, 168, 236, 240
Fluorescent in0sity hybridization (FISH) 215
Fungi 145, 146

G
Garin-Boujadoux-Bannwarth syndrome, 18
Gene therapy, viral 13, 218, 221
Genomic, DNA sequencing 64, 68
Genomic, RNA sequencing 64, 68
Genomic, sequencing 11, 106, 168, 197
Genomic, testing 11, 67, 68, 73, 80, 90, 92
Gout 44
Graves' disease 46
Guidelines, clinical practice 74, 82, 83
Guidelines, treatment 83
Guillain-Barre syndrome 13, 45, 87, 219, 222

H
Hashimoto's thyroiditis 46
Heart block, atrioventricular 10, 12, 22, 53, 138-140-142
Heart disease, rheumatic (strep throat) 87
Heart failure, congestive 54
Heartland & Bourbon virus disease 33-35, 112, 185
Herpes simplex virus, human 39, 61, 69, 131, 143, 144, 146-149
Herpes, zoster 146
H1N1 145
H5N1 145
HIV/AIDS 11, 69, 73, 113, 145-147, 203, 217
Homeopathy 11
Huntington;s disease 217
Hydrogen peroxide, therapy 126
Hyperbaric oxygen, therapy 126
Hypothalamic pituitary adrenal axis (HPA) 49

I
Immune-based treatments 225
Immune system 38-42, 45, 46, 60, 65, 72, 94, 95, 97, 98, 112, 113, 143, 146

Immune system, adaptive 39
Immunization, passive 105
Immunoassay 11
Immunoglobulin G, M 63, 65, 66, 77, 81, 164, 179, 229
Infection, bacterial 39-41, 60
Infection, fungal 39-41, 60, 72
Infection, parasitic 60, 72
Infection, viral 39-41, 60, 72
infection, zoonotic 144
Infectious Diseases of America (IDSA) 74, 80, 192, 229
Infectious load 60
Inflammatory bowel disease 45
Influenza 75
Ispot 11, 64, 67, 73
International Lyme and Associated Diseases Society (ILADS) 8, 10, 11, 13, 36, 41, 74, 81-83, 229, 234-236, 238, 239, 240
Ixodes, cookei 22, 24, 33
Ixodes, hexagonous 18
Ixodes, pacificus 19, 22, 23, 25, 34, 108, 160, 162
Ixodes, persulcatus 22, 27, 108
Ixodes, ricinus 22, 108
Ixodes, scapularis (see also black-legged tick) 7, 18, 19, 22, 23, 25, 27, 28, 33, 35, 101, 103, 106, 108, 110, 160, 162, 170, 172, 178, 180, 216

J
Joints, disease 37, 38
Joints, erosion 57
Joints, inflammation and pain (see also arthritis) 41, 44, 52, 53, 54, 86, 89, 94-98, 121, 123, 124, 138, 206

K
Kaposi sarcoma 148, 149

L
Left-ventricular dysfunction 54
Leukocytes, polymorphonuclear 70
Leukopenia 176
Lupus 11, 44, 69, 73, 113, 140
Lupus, systemic erythematosus 45

Lyme arthritis 57, 59, 61, 64, 66, 67, 72
Lyme carditis 12, 53, 54, 58, 65, 69, 95, 138-142
Lyme Disease Association (LDA) 192
Lyme disease, clinical manifestations 21
Lyme disease, chronic 8, 10, 11, 13, 36-38, 40, 41, 63, 68, 77, 86, 89, 91-98, 112, 143, 144, 162, 207, 229, 230, 239, 240
Lyme disease, chronic complex 36, 38, 86, 91
Lyme disease, chronic syndrome 10, 11, 13, 40, 112, 220
Lyme disease, early-localized 202
Lyme disease, early-disseminated 10, 21, 52, 57-63, 72, 73, 86, 90, 91, 138, 141
Lyme disease, encephalopathy 56, 57
Lyme disease, late-disseminated 10, 21, 36, 55, 57-61, 63, 69, 72, 73, 86, 90, 113
Lyme disease, localized 21, 36, 49, 51, 57, 59-62, 72, 77, 85, 90
Lyme disease, ocular 125
Lyme disease, persistent 55
Lyme disease, post-treatment syndrome 10, 13, 36, 39-41, 63, 86, 87, 91, 112, 162, 198, 215, 216, 221, 230, 236, 239
Lyme disease, transmission 22
Lyme disease, treatment 85-92, 98
Lyme disease, treatment, homeopathic 93
Lymphocytoma, *borrelial* 53
Lymphoma, Hodgkin's 149
Lymphota, Burrito's 149
Lysates 100

M
Magnetic resonance imaging (MRI) 69, 73
Malaise, post-exertional 43
Malaria 150, 179, 180
Malaria, therapy 126
Marijuana, medical 13, 217, 221
Maternal-fetal (transplacental) transmission 225, 226
Measles 131, 145, 147
Memory impairment 61, 88
Meningitis 12, 52, 54, 69, 125, 127, 144, 146-9
Meningitis, aseptic 131
Meningitis, bacterial 132

Meningitis, entero-viral 132
Meningitis, lymphocytic 10, 22, 52-54, 58, 70, 71, 73, 113, 132
Meningitis, mild 54
Meningitis, viral 12, 71, 125, 127, 131, 235
Meningoradiculitis 125
Meningoradiculopathy, lymphocytic 18
Metabolomics 168, 197
Metagenomics 216
Microarray technology 197
Microscopy, dark field 11, 64, 67, 72
Mismacine, therapy 126
Molecular mimicry 219
Mononeuritis, multiplex 52, 54
Mononucleosis, infectious 168
Motor neuron disease 135
Multiple sclerosis 11, 12, 31, 46, 113, 125, 130, 132, 144, 147, 217, 218
Mumps 131
Myalgia 9, 21, 176, 180, 184-187, 236, 240
Mycoplasma, co-infection 161
Myocarditis 10, 22
Myasthenia gravis 46, 135
Myelitis, ataxia 12
Myelitis, osteo 44

N
Nervous system, central 31, 39, 54, 65, 87, 130, 133, 138, 145, 146, 186, 199, 207
Nervous system, peripheral 69, 90, 92, 138, 213
Neuritis, cranial 52, 54
Neuroborreliosis 10, 52, 53, 65, 67-69, 125-127, 143, 144, 148, 200
Neurodegenerative diseases 10, 41, 69, 73, 112, 113, 144, 147-149
Neurofibrillary tangles 127
Neuro-optic pain 240
Neuropathy 217, 219
Neuropathy, cranial 10, 22
Neuropathy, HIV 218
Neuropathy, peripheral 12, 13, 55, 57, 58, 68, 112, 206, 207, 213, 217, 219, 221, 222

O
Octopamine, neurotransmitter 38
Organ, transplant 143
Ornithodoros Spp 24
Outer surface protein 64, 67, 91, 99, 108, 110

P
Palsy, Bell's 12, 71, 126, 127, 133-135
Palsy, facial 10, 22, 52-58, 61, 65. 69-73, 112, 113, 187, 199, 200
Panencephalitis, subacute sclerosing 145, 151
Papilledema 12, 125
Paralysis, partial 161
Parasites 90, 92-96, 98, 145, 146, 180
Paresis 181
Parkinson's disease 12, 31, 38, 144, 145, 147, 148, 217, 218
Parkinsonian movement disorder 54
Parkinsonian syndrome 52
Pathogens 143-144
Periodontitis 168
Phagocytosis, lymphocytic 179
Plague 20
Plaque 127
Plasmapheresis 46, 210
Plasmids 39, 68
Pleocytosis, lymphocytic 68, 90, 92
Pneumonia 143, 149
Pneumonia, chlamydia 37
Pneumonia, mycoplasma 37
Polymerase chain reaction 11, 61, 64, 67, 68, 73, 126, 164, 178, 179, 216, 217, 231, 232
Polyneuropathy 57, 58, 112
Polyneuropathy, chronic axonal 10, 21
Polyneuropathy, chronic inflammatory demyelinating 46
Powassan virus disease 12, 23, 24, 33, 34, 35, 37, 109, 161, 170, 179-182
Prevention 110, 111, 113, 114, 123, 168, 169
Prions 146
Prognosis 110, 112, 115
Protein, amyloid beta 127
Protein, tau 127
Proteomics 197
Protozoa 145, 146

Psoriasis 46
Psychosis 56-58, 112, 113

R

Radiculoneuritis 10, 22
Radiculopathy 52-54, 58, 70, 71, 73, 113
Ramsey-Hunt syndrome 135
Rash, *erythema migrans* 7, 9, 18, 21, 36, 50, 52, 53, 57, 58, 61, 62, 70-72, 74, 82, 83
Reiter's syndrome 87
Respiratory tract infection 35
Ricephalus sanguineus 24, 34, 35
Rickettsiosis, parkeri rickettsia 24, 33, 34, 187
Rickettsiosis, philippi 112
Rickettsiosis, rickettsia 24, 34, 177
Reservoir, infectious 102, 106

S

Schizophrenia 56
Sciatica 70
Sclerosis, amyotrophic lateral 31
Sclerosis, multiple 38, 45, 56, 58, 69, 73
Scrambler therapy 13, 219, 222
Sequencing, next generation 68, 73, 198
Sequencing, whole genome 216
Shingles 70, 71, 73, 113
Single photon emission computed tomography (SPECT) 69, 73
Sjogren's syndrome 47
Southern tick-associated rash illness (STARI) 10, 34, 51, 112, 121, 124
Spider web 70, 71, 73, 113
Spirochete 7, 9, 13
STARI (see Southern tick-associated rash illness)
Stem cells, hematopoietic 74, 161, 162, 168
Stem cells, therapy 210, 218, 222
Surveillance 160, 165-168, 238
Syphilis 145, 168

T

Tendinitis 44
Thrombocytopenia 178
Tick, American dog 24
Tick, black-legged (see also *Ixodes scapularis*) 3, 19, 22, 23, 25, 27, 35, 76, 161, 170, 172, 173, 176, 180, 216, 235, 237, 240
Tick-borne diseases 192, 224, 225, 227
Tick, brown hog 34
Tick, golf coast 24
Tick, ground hog (or woodchuck) 34
Tick, killer, permethrin 110, 111, 123, 124, 167, 207
Tick, lone star (see also *Anaplasma americanum*) 24, 33, 35, 161, 177
Tick, pesticide (acaricides) 117
Tick, repellent, DEET 114, 123, 124
Tick, repellent, IR535 114
Tick, repellent, Nootkatone 167
Tick, repellent, oil of lemon eucalyptus (OLE) 114
Tick, repellent, Picardin 114
Tick, resistance 109
Tick, soft 24, 35
Tick, Western black-legged 19, 22, 23, 25
Tick, wood 34
Tick-borne diseases 8, 9, 223, 225, 230, 239-41
Tick-borne diseases, non-Lyme 217
Ticks, of the U.S. 22
Toxins 95, 113
Toxins, bio 60, 72, 112, 113
Toxins, endo 60, 72, 90, 92, 93, 112, 113
Toxins, myco 60, 112, 113
Toxins, neuro 60, 72, 86, 112, 113
Toxoplasmosis 150
Transplantation, solid organ 74
Tremors 61
Trials, randomized controlled clinical 82, 197, 218, 219
Tularemia (Tularemia, francisella) 20, 24, 34, 112, 161, 182, 187, 190, 191

V

Vaccination, oral-baited 108
Vaccine 8, 11, 13, 99, 168, 196, 203, 225, 233
Vaccine, canine, LymeVax (Fort Dodge Laboratories) 100
Vaccine, canine, Galaxy Lyme (Intervet-Schering-Plough) 100
Vaccine, canine, recombinant (Merial) 100

Vaccine, development 100, 101, 107
Vaccine, development, rodents 103
Vaccine, development, rodents, white-footed mouse (U.S. Biologic and Connecticut Agricultural Experiment Station) 101
Vaccine, human, incorporating antigen against co-infection anaplasmosis 104
Vaccine, human, incorporating antigens 106
Vaccine, human, interruption 105
Vaccine, human, Lymerix (Glaxo-Smith-Kline) 99, 100
Vaccine, human, modifying canine vaccine 104, 106
Vaccine, human, oral bait 106
Vaccine, human, reservoir-targeted 104, 106-108
Vaccine, human, rice-based 103, 106
Vaccine, human, targeting tick saliva 104, 106
Vaccine, human, Vaccinii virus 103, 106, 108
Vaccine, human, vector-targeted, colonization 103
Vaccine, human, vector-targeted, transmission 103, 105, 107-109
Vaccine, human, whole-cell lysates 101, 105
Vaccine, oral bait 103
Vaccine, recombination 99
Vaccine, reservoir-based 101, 103
Vaccine, rodent-targeted 216
Valvular heart disease 138
Varicella 131
Vasculitis 46
Vector-borne diseases 223, 224, 227, 230
Vertigo 56, 58, 176, 180, 181

W

West Nile virus disease 131, 146, 149
Western immunoblot (see Assays)

X

Xenodiagnosis 198
Xenodiagnostic marker 101, 217

Z

Zoonosis 7, 9, 76

Author Index

A
Aberer E 246
Accardo S 243
Afzelius A 18
Agger WA 242
Agre F 242
Aguero-Rosenfeld ME 242
Albert S 242
Alder J 242
Al-Robaiy S 242
Altman DG
Archer H 247
Asch ES 242
Aucott JN 242
Auwaerter PG 242
Aversa JM 249
Aylward A 246

B
Bacon RM 242
Badiaga S 243
Bannwarth A 18
Barbour AG 242, 251
Barsic B 242
Bartenhagen NH 250
Barthold SW 242, 246
Bastian SL 249
Bell C 134, 135
Belongia EA 250
Benach JL 250
Berger BW 243
Bitter B 130
Bittker S 242
Blossom B 244
Borg R 243
Borrel A 7
Bowers LC 247

Brade V 242
Bradley JF 243
Brandt KS 251
Britton CB 244
Brouqui P 243
Brown 108
Buchwell A 18
Bujadoux 18
Bujak DI 242
Burgdorfer W 18, 40, 227
Burckhardt CS 246

C
Cabello FC 243
Cairns V 243
Calkins H 250
Callister SM 242
Cameron DJ 243
Cartter ML 239
Cerar D 243
Cerar T 243
Chandran A 249
Charcot JM 132-135
Childs JA 244
Cimmino MA 243
Cimperman J 248, 250
Clark RP 243
Cohen TJ 247
Coleman 107
Collins JJ 246
Comstedt P 102, 104, 108
Conaty SM
Cooper C 243
Corapi KM 243
Corbera KM 244
Costello CM 243
Coyle PK 243, 247
Craft JE 250
Crowder LA 242
Cunha BA 243

D
Dadey E 251
Dattwyler RJ 243, 244, 247, 251

Davidson J 96
Delong AK 244
Dennis DT 248
Deodhar AA 246
Dihazi H 242
Dobrikova EY 249
Dolan MC 248, 251
Donta ST 244
Dotevall L 243, 244
Dressler F 229, 248
Duray PH 244

E
Eikeland R 244
Embers ME 244
Eppes SC 244
Ernerudh J 249
Evans J 247

F
Fallon BA 215, 220, 244
Feder HM Jr 243
Feng J 216
Feng S 246
Fikrig 107, 108
Fish D 248
Fix AD 244
Francois LK 250
Frank C, 244
Frank E 248
Freet KJ 246
Fried M 220
Fymat AL 1, 3, 146, 147, 244-246

G
Gaillard F 135
Garin 18
Gerber MA 246, 249
Glimaker K 249
Gilmore RD 107, 247
Godfrey HP 243
Godwin J 243
Goodman JL 243
Green J 250

Green CL 250
Grimson R 247
Grunwaldt E 244

H
Hardham JM 251
Hagberg L 243, 244
Hagman 108
Hahn G 246
Halperin JJ 243
Hanincova K 248
Hanson 108
Happ CM 248
Hardesty D
Hartiala P 246
Haupl T 246
Hayes E 246
Herlofson K 244
Hiekkanen H 248
Hodzic E 242, 246
Hojgaard A 248
Holabird NB 249
Holden K 246
Hornik A 130
Hu LT 243, 247
Hufstader M 249
Hunfeld KP 246
Hunziker T 249
Hutchinson GJ 250
Hyman LG 247
Hytonen J 246

I
Imai DM 242
Ito Y 244

J
Jacobs MB 247
Jarvis K 242
Jobe D 242
Johnson L 246, 250
Johnson RC 243, 246, 247
Johnson SM 250
Jones KD 246

Jurkovich PJ 246

K
Kacza J 242
Kalp J 215
Kaplan RF 247
Kaye KM 149
Keilp JG 244
Kersten A 246
Kirschfink M 247
Klein D 249
Klempner MS 246, 247
Kluger DM 247
Kodner CB 246
Kortte KB 242
Kraiczy P 247
Krause PJ 247
Krupp LB 247, 248
Kugeler KJ 242
Kunkel MJ 244

L
Lawrence C 247
Lecomte P 247
Ledel A 249
Lee DH 250
Lee J 247
Lee-Lewandowsky 217
Lefebvre 107
Levin ES 244
Leyden JJ 249
Liang FT 247
Liegner KB 247
Lipton RB 247
Livengood JA 247
Ljostad U 244
Lo R, Menzies DJ 247
Logar M 247
Logigian EL 247, 249, 250
Lotric-Furlan S 248, 250
Lowy FD 247
Luft BJ 243, 244, 247
Luger SW 247, 248
Lunetta JM 250

M
Maes E 247
Majerus L 242
Maki DG 247
Malawista SE 249
Maloney EL 244, 248
Mankoff J 246
Maraspin V 247, 248
Maretic T 242
Marjamaki M 248
Maraspin V 250
Marques A 248
Massarotti EM 248
Massung RF 251
Matter L 251
McKenna D 248
Mead PS 242
Meltzer MI 248
Middelveen M 249
Millner MM 248
Mitten M 242
Moesli P 249
Moggiana GL 243
Morrison C 242
Mpoy CM 249
Mukherjee P 248
Mullegger RR 248
Munoz B 242
Mursic VP 248
Mygland A 244

N
Nadelman RB 248, 250
Neitzel D 250
Neubert U 250
Newman SA 243, 249
Nikoskelainen J 248
Nocton JJ 248
Nordwall M 249
Noring R 246
Norris SJ 251
Nowakowski J 242, 248, 250, 251

O
Oksi J 248
Orloski KA 248

P
Paparone P 247
Parisi M 243
Pass H 243
Petkova E 244
Pena CA 244
Petri M 250
Pfister HW 248, 249
Philipp MT 107, 244, 247
Phillips CB 249
Phillips SE 243, 24
Piesman J 248, 249
Pinkerton RE 243
Poitschek C 246
Pollina DA 248
Preac-Mursic V 248, 249, 250, 251
Probert 107

R
Rahn DW 248, 249, 250
Ramamoorthy R 244
Ramanathan R 251
Ramsay D 247
Raoult D 243
Rauch S 246
Raveche ES 220
Ray N 247
Rebman AW 242
Reed KD 250
Reinhardt S 248
Reiter M 249
Restrepo BI 242
Riegel H 242
Rittig M 246
Rodriguez-Porcel S 130
Rogers RA 246
Rosenblum J 130
Rutledge BJ 248

Ruzic E 250
Ruzic-Sabljic E 243, 247

S
Sapi I 216, 249
Sartakova ML 249
Savely VR 250
Schaad UB 249
Schaefer C 249
Schmidli J 249
Schoen RT 249, 250
Schulze J 242
Schutzer SE 243
Schwan TG 249
Schwartz R 242
Schweitzer PJ 244
Seltzer EG 249
Shadick NA 249
Shapiro ED 246, 249-251
Shapiro JR 247
Sharma B 216
Sjowall J 249
Skerka C 247
Skogman BH 249
Sliwinski M 248
Smith K 249
Sparrow SS 250
Sperling J 249
Spiegel H 249
Spielman A 247
Spork KD 248
Squires NK 248
Stanek G 248, 249
Steere AC 243, 247, 249, 250
Stricker RB 246, 250
Strickland GT 244
Strle F 250
Strugar J 242
Suhonen J 246
Swanson SJ 250
Szczepanski A 250

T
Tager A 220
Tang S 250
Telford SR 3rd 107, 247, 248
Terekhova DA 249
Thaisetthawatkul P 250
Thompson GR 3rd 250
Thurmayr R 250, 251
Turk SP 248

V
Vazquez M 250
Viljanen MK 248
Volkman DJ 243, 244, 250

W
Wanner G 248
Warshafsky S 250
Watts T 96
Weber K 249-251
Weder B 251
Weigand J 246
Weiss M 242
Wellman J 134
White MI 243
Wichelhaus TA 246
Wiedersheim P 251
Wilcox S 246
Wilske B 248, 251
Wormser GP 247 251

Y
Yin SR 244

Z
Zeidner NS 251
Zemel LS 246
Zhang JR 251

DR. ALAIN L. FYMAT is a medical-physical scientist and an educator. He is the current President/CEO and Institute Professor at the International Institute of Medicine & Science with a previous appointment as Executive Vice President/Chief Operating Officer and Professor at the Weil Institute of Critical Care Medicine, California, U.S.A. He was formerly Professor of Radiology, Radiological Sciences, Radiation Oncology, Critical Care Medicine, and Physics at several U.S. and European Universities. Earlier, he was Deputy Director (Western Region) of the U.S. Department of Veterans Affairs (Office of Research Oversight). At the Loma Linda Veterans Affairs Medical Center, he was Scientific Director of Radiology, Director of the Magnetic Resonance Imaging Center and, for a time, Acting Chair of Radiology. Previously, he was Director of the Division of Biomedical and Biobehavioral Research at the University of California at Los Angeles/Drew University of Medicine and Science. He was also Scientific Advisor to the U.S. National Academy of Sciences, National Research Council, for its postdoctoral programs tenable at the California Institute of Technology and Member of the Advisory Group for Research & Development, North Atlantic Treaty Organization (NATO). He is Health Advisor to the American Heart & Stroke Association, Coachella Valley Division, California. He is a frequent Keynote Speaker and Organizing Committee member at several international scientific/medical conferences. He has lectured extensively in the U.S.A, Canada, Europe, Asia, and Africa. He has published ~ 475 scholarly scientific publications and books. He is also Editor-in-Chief, Honorable Editor or Editor of numerous medical/scientific Journals to which he regularly contributes. He is a member of the New York Academy of Sciences and the European Union Academy of Sciences, a Board member of several institutions, and a reviewer for the prestigious UNESCO Newton Prize, United Kingdom National Commission for UNESCO.

Dr. Fymat's current research interests are focused on neurodegenerative diseases (Alzheimer's, Parkinson's, epilepsy, dementias, and others), oncology (glioblastoma), epigenetics & ecogenetics, and nanomedicine & nanobiotechnology. These are represented in part in his latest books: "From the heart to the brain: my collected works in medical science research (2016-2018)", "**Alzhei ...Who?** Demystifying the disease and what you can do about it", "**The Odyssey of Humanity's Diseases**: Epigenetic and ecogenetic modulations from ancestry through inheritance, environment, culture and behavior" Volumes 1, 2, and 3, and "***Parkin..ss..oo..nn***: Elucidating the disease and what you can do about it". Having previously written several book chapters on malaria, Ebola, and Zika, this is his first in-depth study of the emerging Lyme disease and its co- and secondary infections.